大数据与人工智能技术丛书

人工智能专业英语教程

English for Artificial Intelligence

◎ 张强华 司爱侠 编著

清华大学出版社

北京

内容简介

本书是人工智能专业英语教材,涉及人工智能基础、强人工智能与弱人工智能、人工智能应用领域、常用搜索算法、软件与 Python 编程语言、知识系统与知识库、专家系统和推理引擎、机器学习与深度学习、决策树、人工神经网络和模糊神经网络、监督学习与无监督学习、人工智能安全、大数据与人工智能、云计算与人工智能、普适计算、人工智能与智慧城市等多个方面。

本书力求在体例上有创新,适合教学。每个单元均包含以下部分:课文——选材广泛、风格多样、切合实际的两篇专业文章;单词——给出课文中出现的新词,读者由此可以积累人工智能专业的基本词汇;词组——给出课文中的常用词组;缩略语——给出课文中出现的、业内人士必须掌握的缩略语;习题——既有针对课文的练习,也有词汇练习和专业性短文填空;专业短文翻译——培养读者的翻译能力;参考译文——让读者对照理解和提高翻译能力。

本书吸纳了作者二十余年的 IT 行业英语翻译与图书编写经验,与课堂教学的各个环节紧密切合,支持备课、教学、复习及考试各个教学环节,有配套的教学课件、参考答案等。

本书既可作为高等院校人工智能相关专业的专业英语教材,也可供相关从业人员自学,作为培训班教材,亦颇得当。

图书在版编目(CIP)数据

人工智能专业英语教程=English for Artificial Intelligence/张强华,司爱侠编著.—北京:清华大学出版社,2021.5
(大数据与人工智能技术丛书)
ISBN 978-7-302-55886-6

Ⅰ. ①人… Ⅱ. ①张… ②司… Ⅲ. ①人工智能-英语-高等学校-教材 Ⅳ. ①TP18

中国版本图书馆 CIP 数据核字(2020)第 111360 号

策划编辑:魏江江
责任编辑:王冰飞
封面设计:刘 键
责任校对:时翠兰
责任印制:宋 林

出版发行:清华大学出版社
　　　网　　址:http://www.tup.com.cn,http://www.wqbook.com
　　　地　　址:北京清华大学学研大厦 A 座　　　　　邮　　编:100084
　　　社 总 机:010-62770175　　　　　　　　　　 邮　　购:010-83470235
　　　投稿与读者服务:010-62776969,c-service@tup.tsinghua.edu.cn
　　　质量反馈:010-62772015,zhiliang@tup.tsinghua.edu.cn
　　　课件下载:http://www.tup.com.cn,010-83470236
印 装 者:三河市龙大印装有限公司
经　　销:全国新华书店
开　　本:185mm×260mm　　　印　　张:16.25　　　　字　　数:376 千字
版　　次:2021 年 5 月第 1 版　　　　　　　　　　　印　　次:2021 年 5 月第 1 次印刷
印　　数:1～1500
定　　价:49.80 元

产品编号:085425-01

前　言

我们正在进入人工智能时代。我国人工智能技术已经进入爆发期，许多高校都开设了人工智能专业，培养专业人才。由于人工智能发展速度很快，从业人员必须掌握许多新技术、新方法，因此对专业英语要求较高。具备相关职业技能并精通专业英语的人员往往会赢得竞争优势，成为职场中不可或缺的核心骨干与领军人才。本书就是为提高行业人才专业英语水平而编写的。

本书的特点与优势如下。

（1）选材全面，涉及人工智能基础、强人工智能与弱人工智能、人工智能应用领域、常用搜索算法、软件与 Python 编程语言、知识系统与知识库、专家系统和推理引擎、机器学习与深度学习、决策树、人工神经网络和模糊神经网络、监督学习与无监督学习、人工智能安全、大数据与人工智能、云计算与人工智能、普适计算、人工智能与智慧城市等各个方面。书中许多内容都非常实用，具有广泛的覆盖面。

（2）体例创新，对课文选材进行了严谨推敲与细致加工，使其具有教材特性。与课堂教学的各个环节紧密切合，支持备课、教学、复习及考试各个教学环节。每个单元均包含以下部分：课文——选材广泛、风格多样、切合实际的专业文章；单词——给出课文中出现的新词，读者由此可以积累人工智能专业的基本词汇；词组——给出课文中的常用词组；缩略语——给出课文中出现的、业内人士必须掌握的缩略语；习题——既有针对课文的练习，也有词汇练习和专业性短文填空；专业短文翻译——培养读者的翻译能力；参考译文——让读者对照理解和提高翻译能力。附录中的"词汇表"收集了全书课文的词汇，读者既可用它背诵词汇，也可将它作为专业小词典长期查阅。

（3）习题量适当，题型丰富，难易搭配，便于教师组织教学。

（4）教学资源完善，有配套的教学大纲、教学课件、参考答案等，可扫描封底的"书圈"二维码，在公众号"书圈"中申请下载。

（5）作者为每一单元的 A、B 两篇课文录制了音频，扫描章首的二维码即可收听。

（6）作者有二十余年 IT 行业英语图书的编写经验。在作者编写的专业英语书籍中，有三部国家级"十一五"规划教材、一部全国畅销书和一部获"华东地区教材二等奖"的图书。编写这些图书的经验有助于本书的完善与提升。

本书既可作为高等院校人工智能相关专业的专业英语教材，也可供相关从业人员自学，作为培训班教材，亦颇得当。

由于作者学识有限，本书定有一些不妥之处，望大家不吝赐教，让我们共同努力，使本书成为一部"符合学生实际、切合行业实况、知识实用丰富、严谨开放创新"的优秀教材。

作　者

2021 年 1 月

目　录

Unit 1

录音

Text A

Artificial Intelligence（AI）

Artificial intelligence（AI）is the simulation of human intelligence processes by machines，especially computer systems. These processes include learning（the acquisition of information and rules for using the information），reasoning（using rules to reach approximate or definite conclusions）and self-correction. Particular applications of AI include expert systems，speech recognition and machine vision.

AI can be categorized as either weak or strong. Weak AI，also known as narrow AI，is an AI system that is designed and trained for a particular task. Virtual personal assistants，such as Apple's Siri，are a form of weak AI. Strong AI，also known as artificial general intelligence，is an AI system with generalized human cognitive abilities. When presented with an unfamiliar task，a strong AI system is able to find a solution without human intervention.

Because hardware，software and staffing costs for AI can be expensive，many vendors are including AI components in their standard offerings，as well as access to artificial intelligence as a service（AIaaS）platforms. AI as a service allows individuals and companies to experiment with AI for various business purposes and sample multiple platforms before making a commitment. Popular AI cloud offerings include

Amazon AI services，IBM Watson Assistant，Microsoft Cognitive Services and Google AI services.

While AI tools present a range of new functionality for businesses，the use of artificial intelligence raises ethical questions. This is because deep learning algorithms，which underpin many of the most advanced AI tools，are only as smart as the data they are given in training. Because a human selects what data should be used for training an AI program，the potential for human bias is inherent and must be monitored closely.

Some industry experts believe that the term artificial intelligence is too closely linked to popular culture，causing the general public to have unrealistic fears about artificial intelligence and improbable expectations about how it will change the workplace and life in general. Researchers and marketers hope augmented intelligence，which has a more neutral connotation，will help people understand that AI will simply improve products and services and will not replace the humans that use them.

1．Types of AI

Arend Hintze，an assistant professor of integrative biology and computer science and engineering at Michigan State University，categorizes AI into four types，from the kind of AI systems that exist today to sentient systems which do not yet exist. His categories are as follows.

（1）Type 1：Reactive machines. An example is Deep Blue，the IBM chess program that beat Garry Kasparov in the 1990s. Deep Blue can identify pieces on the chess board and make predictions，but it has no memory and cannot use past experiences to inform future ones. It analyzes possible moves—its own and its opponent—and chooses the most strategic move. Deep Blue and Google's AlphaGo were designed for narrow purposes and cannot easily be applied to another situation.

（2）Type 2：Limited memory. These AI systems can use past experiences to inform future decisions. Some of the decision-making functions in self-driving cars are designed this way. Observations inform actions happening in the not-so-distant future，such as a car changing lanes. These observations are not stored permanently.

（3）Type 3：Theory of mind. This psychology term refers to the understanding that others have their own beliefs，desires and intentions that impact the decisions they make. This kind of AI does not yet exist.

（4）Type 4：Self-awareness. In this category，AI systems have a sense of self，have consciousness. Machines with self-awareness understand their current state and can use the information to infer what others are feeling. This type of AI does not yet exist.

2. Examples of AI technology

AI is incorporated into a variety of different types of technology. Here are some examples.

(1) Automation: What makes a system or process function automatically. For example, robotic process automation (RPA) can be programmed to perform high-volume, repeatable tasks that humans normally performed. RPA is different from IT automation in that it can adapt to changing circumstances.

(2) Machine learning: The science of getting a computer to act without programming. Deep learning is a subset of machine learning that, in very simple terms, can be thought of as the automation of predictive analytics. There are three types of machine learning algorithms:

- Supervised learning: Data sets are labeled so that patterns can be detected and used to label new data sets.
- Unsupervised learning: Data sets aren't labeled and are sorted according to similarities or differences.
- Reinforcement learning: Data sets aren't labeled, but after performing an action or several actions, the AI system is given feedback.

(3) Machine vision: The science of allowing computers to see. This technology captures and analyzes visual information using a camera, analog-to-digital conversion and digital signal processing. It is often compared to human eyesight, but machine vision isn't bound by biology and can be programmed to see through walls, for example. It is used in a range of applications from signature identification to medical image analysis. Computer vision, which is focused on machine-based image processing, is often conflated with machine vision.

(4) Natural language processing (NLP): The processing of human—and not computer—language by a computer program. One of the older and best known examples of NLP is spam detection, which looks at the subject line and the text of an email and decides if it's junk. Current approaches to NLP are based on machine learning. NLP tasks include text translation, sentiment analysis and speech recognition.

(5) Robotics: A field of engineering focused on the design and manufacturing of robots. Robots are often used to perform tasks that are difficult for humans to perform or perform consistently. They are used in assembly lines for car production or by NASA to move large objects in space. Researchers are also using machine learning to build robots that can interact in social settings.

（6）Self-driving cars：These use a combination of computer vision，image recognition and deep learning to build automated skill at piloting a vehicle while staying in a given lane and avoiding unexpected obstructions，such as pedestrians.

3. AI applications

AI has made its way into a number of areas. Here are six examples.

（1）AI in healthcare. The biggest bets are on improving patient outcomes and reducing costs. Companies are applying machine learning to make better and faster diagnoses than humans. One of the best known healthcare technologies is IBM Watson. It understands natural language and is capable of responding to questions asked of it. The system mines patient data and other available data sources to form a hypothesis，which then presents with a confidence scoring schema. Other AI applications include chatbots，a computer program used online to answer questions and assist customers，to help schedule follow-up appointments or aid patients through the billing process，and virtual health assistants that provide basic medical feedback.

（2）AI in business. Robotic process automation is being applied to highly repetitive tasks normally performed by humans. Machine learning algorithms are being integrated into analytics and CRM platforms to uncover information on how to better serve customers. Chatbots have been incorporated into websites to provide immediate service to customers. Automation of job positions has also become a talking point among academics and IT analysts.

（3）AI in education. AI can automate grading，giving educators more time. AI can assess students and adapt to their needs，helping them work at their own pace. AI tutors can provide additional support to students，ensuring they stay on track. AI could change where and how students learn，perhaps even replacing some teachers.

（4）AI in finance. AI in personal finance applications，such as Mint or Turbo Tax，is making a break in financial institutions. Applications such as these collect personal data and provide financial advice. Other programs，such as IBM Watson，have been applied to the process of buying a home. Today，software performs much of the trading on Wall Street.

（5）AI in law. The discovery process，sifting through documents，in law is often overwhelming for humans. Automating this process is a more efficient use of time. Startups are also building question-and-answer computer assistants that can sift programmed-to-answer questions by examining the taxonomy and ontology associated with a database.

（6）AI in manufacturing. This is an area that has been at the forefront of incorporating robots into the workflow. Industrial robots used to perform single tasks and were separated from human workers, but as the technology advanced that changed.

4. Security and ethical concerns

The application of AI in the realm of self-driving cars raises security as well as ethical concerns. Cars can be hacked, and when an autonomous vehicle is involved in an accident, liability is unclear. Autonomous vehicles may also be put in a position where an accident is unavoidable, forcing the programming to make an ethical decision about how to minimize damage.

Another major concern is the potential for abuse of AI tools. Hackers are starting to use sophisticated machine learning tools to gain access to sensitive systems, complicating the issue of security beyond its current state.

Deep learning-based video and audio generation tools also present bad actors with the tools necessary to create so-called deepfakes, convincingly fabricated videos of public figures saying or doing things that never took place.

5. Regulation of AI technology

Despite these potential risks, there are few regulations governing the use of AI tools, and where laws do exist, they typically pertain to AI only indirectly. For example, federal fair lending regulations require financial institutions to explain credit decisions to potential customers, which limit the extent to which lenders can use deep learning algorithms, which by their nature are typically opaque. Europe's GDPR puts strict limits on how enterprises can use consumer data, which impedes the training and functionality of many consumer-orientated AI applications.

In 2016, the National Science and Technology Council (NSTC) issued a report examining the potential role governmental regulation might play in AI development, but it did not recommend specific legislation be considered. Since that time the issue has received little attention from lawmakers.

✎ New Words

simulation	[ˌsɪmjʊˈleɪʃn]	n. 模仿,模拟
acquisition	[ˌækwɪˈzɪʃn]	n. 获得
rule	[ruːl]	n. 规则,规定;统治,支配
		v. 控制,支配
reasoning	[ˈriːzənɪŋ]	n. 推理,论证

		v. 推理,思考;争辩;说服
		adj. 推理的
approximate	[ə'prɒksɪmɪt]	adj. 极相似的
	[ə'prɒksɪmeɪt]	vi. 接近于,近似于
		vt. 靠近,使接近
definite	['defɪnɪt]	adj. 明确的;一定的;肯定
conclusion	[kən'kluːʒn]	n. 结论;断定,决定;推论
self-correction	[ˌselfkə'rekʃn]	n. 自校正;自我纠错;自我改正
particular	[pə'tɪkjʊlə]	adj. 特别的;详细的;独有的
		n. 特色,特点
vision	['vɪʒn]	n. 视觉
narrow	['nærəʊ]	adj. 狭隘的,狭窄的
virtual	['vɜːtʃʊəl]	adj. (计算机)虚拟的;实质上的,事实上的
cognitive	['kɒgnɪtɪv]	adj. 认知的,认识的
ability	[ə'bɪlɪtɪ]	n. 能力,资格;能耐,才能
unfamiliar	[ˌʌnfə'mɪlɪə]	adj. 不熟悉的;不常见的;陌生的;没有经验的
experiment	[ɪk'sperɪmənt]	n. 实验,试验;尝试
		vi. 做实验
commitment	[kə'mɪtmənt]	n. 承诺,许诺;委任,委托
ethical	['eθɪkl]	adj. 道德的,伦理的
underpin	[ˌʌndə'pɪn]	vt. 加固,支撑
unrealistic	[ˌʌnrɪə'lɪstɪk]	adj. 不切实际的;不现实的;空想的
fear	[fɪə]	n. 害怕;可能性
		vt. 害怕;为……忧虑(或担心、焦虑)
		vi. 害怕;忧虑
expectation	[ˌekspek'teɪʃn]	n. 期待;预期
neutral	['njuːtrəl]	adj. 中立的
connotation	[ˌkɒnə'teɪʃn]	n. 内涵,含义
integrative	['ɪntɪgreɪtɪv]	adj. 综合的,一体化的
sentient	['sentɪənt]	adj. 有感觉能力的,有知觉力的
reactive	[rɪ'æktɪv]	adj. 反应的
prediction	[prɪ'dɪkʃn]	n. 预测,预报;预言
opponent	[ə'pəʊnənt]	n. 对手
observation	[ˌɒbzə'veɪʃn]	n. 观察,观察力
psychology	[saɪ'kɒlədʒɪ]	n. 心理学;心理特点;心理状态
intention	[ɪn'tenʃn]	n. 意图,目的;意向
self-awareness	[self-ə'weənɪs]	n. 自我意识
consciousness	['kɒnʃəsnɪs]	n. 意识,观念;知觉

circumstance	[ˈsɜːkəmstəns]	n.环境,境遇
similarity	[ˌsɪmɪˈlærɪtɪ]	n.相像性,相仿性,类似性
signature	[ˈsɪgnɪtʃə]	n.签名;署名;识别标志
identification	[aɪˌdentɪfɪˈkeɪʃn]	n.鉴定,识别
detection	[dɪˈtekʃn]	n.检查,检测
junk	[dʒʌŋk]	vt.丢弃,废弃
		n.废品;假货
consistently	[kənˈsɪstəntlɪ]	adv.一贯地,坚持地
pilot	[ˈpaɪlət]	n.引航员;向导
		vt.驾驶
vehicle	[ˈviːɪkl]	n.车辆;交通工具
unexpected	[ˌʌnɪkˈspektɪd]	adj.意外的;忽然的;突然的
obstruction	[əbˈstrʌkʃn]	n.阻塞,阻碍,受阻
pedestrian	[pəˈdestrɪən]	n.步行者,行人
		adj.徒步的
healthcare	[ˈhelθkeə]	n.卫生保健
diagnose	[ˈdaɪəgnəʊz]	vt.诊断;判断
		vi.做出诊断
hypothesis	[haɪˈpɒθəsɪs]	n.假设,假说;前提
chatbot	[tʃætbɒt]	n.聊天机器人
appointment	[əˈpɔɪntmənt]	n.预约
repetitive	[rɪˈpetɪtɪv]	adj.重复的,啰唆的
overwhelming	[ˌəʊvəˈwelmɪŋ]	adj.势不可挡的,压倒一切的
taxonomy	[tækˈsɒnəmɪ]	n.分类学,分类系统
ontology	[ɒnˈtɒlədʒɪ]	n.本体,存在;实体论
forefront	[ˈfɔːfrʌnt]	n.前列;第一线;活动中心
incorporating	[ɪnˈkɔːpəreɪtɪŋ]	v.融合,包含;使混合
realm	[relm]	n.领域,范围
accident	[ˈæksɪdənt]	n.意外事件;事故
unclear	[ˌʌnˈklɪə]	adj.不清楚的,不明白的,含糊不清
unavoidable	[ˌʌnəˈvɔɪdəbl]	adj.不可避免的,不得已的
minimize	[ˈmɪnɪmaɪz]	vt.把……减至最低数量[程度],最小化
damage	[ˈdæmɪdʒ]	n.损害,损毁;赔偿金
		v.损害,毁坏
deepfake	[ˈdiːpfeɪk]	n.换脸术
convincingly	[kənˈvɪnsɪŋlɪ]	adv.令人信服地,有说服力地
fabricate	[ˈfæbrɪkeɪt]	vt.编造,捏造
regulation	[ˌregjʊˈleɪʃn]	n.规章,规则

		adj. 规定的
credit	[ˈkredɪt]	*n.* 信誉，信用；［金融］贷款
		vt. 相信，信任
opaque	[əʊˈpeɪk]	*adj.* 不透明的；含糊的
		n. 不透明
strict	[strɪkt]	*adj.* 严格的；精确的；绝对的
impede	[ɪmˈpiːd]	*vt.* 阻碍；妨碍；阻止
lawmaker	[ˈlɔːmeɪkə]	*n.* 立法者

✎ Phrases

human intelligence	人类智能
expert system	专家系统
speech recognition	语音识别
machine vision	机器视觉
be categorized as…	被分类为……
weak AI	弱人工智能
virtual personal assistant	虚拟个人助理
strong AI	强人工智能
artificial general intelligence	通用人工智能
for…purpose	为了……目的
a range of	一系列，一些，一套
deep learning algorithm	深度学习算法
computer science	计算机科学
sentient system	感觉系统
self-driving car	自动驾驶汽车
not-so-distant future	不远的将来
a sense of…	一种……感觉
be incorporated into…	被并入……
predictive analytic	预测分析
supervised learning	有监督学习
unsupervised learning	无监督学习
reinforcement learning	强化学习
analog-to-digital conversion	模（拟）数（字）转换
digital signal	数字信号
medical image analysis	医学图像分析
machine-based image processing	基于机器的图像处理
be conflated with…	与……混为一谈

spam detection	垃圾邮件检测
text translation	文本翻译
sentiment analysis	情感分析,倾向性分析
assembly line	(工厂产品的)装配线,流水线
social setting	社会环境,社会场景,社会情境
image recognition	图像识别
confidence scoring schema	置信评分模式
virtual health assistant	虚拟健康助理
talking point	话题;论题;论据
financial institution	金融机构

✍ Abbreviations

AI (Artificial Intelligence)	人工智能
AIaaS (Artificial Intelligence as a Service)	人工智能即服务
RPA (Robotic Process Automation)	机器人流程自动化
NLP (Natural Language Processing)	自然语言处理
NASA (National Aeronautics and Space Administration)	美国航空航天局
CRM (Customer Relationship Management)	客户关系管理
GDPR (General Data Protection Regulation)	普通数据保护条例
NSTC (National Science and Technology Council)	国家科学技术委员会

✍ Exercises

【Ex.1】 根据课文内容回答问题。

1) What is artificial intelligence? What do these processes include?

2) What do particular applications of AI include?

3) What can AI be categorized as? What are they respectively?

4) What does AI as a service allow individuals and companies to do?

5) What do researchers and marketers hope the label augmented intelligence will do?

6) How many types does Arend Hintze categorize AI into? What are they?

7) AI is incorporated into a variety of different types of technology. What are some examples mentioned in the passage?

8) How many types of machine learning algorithms are there? What are they?

9) What are the areas AI has made its way into?

10) What did the National Science and Technology Council (NSTC) do in 2016?

【Ex.2】 把下列单词或词组中英互译。

1. *n*.预测,预报;预言 _____　　　 1. _____

2. 有监督学习　　　　　　　　　2. ＿＿＿＿＿＿＿＿＿＿＿＿

3. *n*.规章,规则　*adj*.规定的　　3. ＿＿＿＿＿＿＿＿＿＿＿＿

4. *adj*.认知的,认识的　　　　　4. ＿＿＿＿＿＿＿＿＿＿＿＿

5. *n*.意识,观念；知觉　　　　　5. ＿＿＿＿＿＿＿＿＿＿＿＿

6. signature　　　　　　　　　　6. ＿＿＿＿＿＿＿＿＿＿＿＿

7. artificial general intelligence　7. ＿＿＿＿＿＿＿＿＿＿＿＿

8. reinforcement learning　　　　8. ＿＿＿＿＿＿＿＿＿＿＿＿

9. predictive analytic　　　　　　9. ＿＿＿＿＿＿＿＿＿＿＿＿

10. speech recognition　　　　　10. ＿＿＿＿＿＿＿＿＿＿＿＿

【Ex.3】　短文翻译。

Strong Artificial Intelligence（Strong AI）

Strong artificial intelligence（strong AI）is an artificial intelligence construct that has mental capabilities and functions that mimic the human brain. In the philosophy of strong AI，there is no essential difference between the piece of software，which is the AI，exactly emulating the actions of the human brain，and actions of a human being，including its power of understanding and even its consciousness.

Strong artificial intelligence is also known as full AI.

Strong artificial intelligence is more of a philosophy rather than an actual approach to creating AI. It is a different perception of AI wherein it equates AI to humans. It stipulates that a computer can be programmed to actually be a human mind，to be intelligent in every sense of the word，to have perception，beliefs and have other cognitive states that are normally only ascribed to humans.

However，since humans cannot even properly define what intelligence is，it is very difficult to give a clear criterion as to what would count as a success in the development of strong artificial intelligence. Weak AI，on the other hand，is very achievable because of how it stipulates what intelligence is. Rather than try to fully emulate a human mind，weak AI focuses on developing intelligence concerned with a particular task or field of study. That is a set of activities that can be broken down into smaller processes and therefore can be achieved in the scale that is set for it.

【Ex.4】　将下列词填入适当的位置（每词只用一次）。

narrow	intelligence	simulation	powerful	human
learning	rules	conversations	cognitive	artificial

Weak Artificial Intelligence

Weak artificial intelligence（weak AI）is an approach to artificial intelligence research and development with the consideration that AI is and will always be a __1__ of human cognitive function, and that computers can only appear to think but are not actually conscious in any sense of the word. Weak AI simply acts upon and is bound by the __2__ imposed on it and it could not go beyond those rules. A good example of weak AI is characters in a computer game that act believably within the context of their game character, but are unable to do anything beyond that.

Weak artificial intelligence is also known as narrow __3__ intelligence.

Weak artificial intelligence is a form of AI specifically designed to be focused on a __4__ task and to seem very intelligent at it. It contrasts with strong AI, in which an AI is capable of all and any __5__ functions that a human may have, and is in essence no different than a real human mind. Weak AI is never taken as a general __6__ but rather a construct designed to be intelligent in the narrow task that it is assigned to.

A very good example of a weak AI is Apple's Siri, which has the Internet behind it serving as a __7__ database. Siri seems very intelligent, as it is able to hold a conversation with actual people, even giving snide remarks and a few jokes, but actually operates in a very narrow, predefined manner. However, the 'narrowness' of its function can be evidenced by its inaccurate results when it is engaged in __8__ that it is not programmed to respond to.

Robots used in the manufacturing process can also seem very intelligent because of the accuracy and the fact that they are doing very complicated actions that could seem incomprehensible to a normal __9__ mind. But that is the extent of their intelligence; they know what to do in the situations that they are programmed for, and outside of that they have no way of determining what to do. Even AI equipped for machine __10__ can only learn and apply what it learns to the scope it is programmed for.

Text B

Top 10 Hot Artificial Intelligence (AI) Technologies

The market for artificial intelligence (AI) technologies is flourishing. Beyond the hype and the heightened media attention, the numerous startups and the internet giants racing to acquire them, there is a significant increase in investment and adoption by enterprises. A Narrative Science survey found last year that 38% of enterprises are already using AI, growing to 62% by 2018. Forrester Research predicted a greater than 300% increase in investment in artificial intelligence in 2017 compared with 2016. IDC estimated that the AI market will grow from $8 billion in 2016 to more than $47 billion in 2020.

Coined in 1955 to describe a new computer science sub-discipline, Artificial intelligence today includes a variety of technologies and tools, some time-tested, others relatively new. To help make sense of what's hot and what's not, Forrester just published a TechRadar report on artificial intelligence (for application development professionals), a detailed analysis of 13 technologies enterprises should consider adopting to support human decision-making.

Based on Forrester's analysis, here's my list of the 10 hottest AI technologies.

1. Natural Language Generation

Producing text from computer data. Currently used in customer service, report generation, and summarizing business intelligence insights. Sample vendors: Attivio, Automated Insights, Cambridge Semantics, Digital Reasoning, Lucidworks, Narrative Science, SAS, Yseop.

2. Speech Recognition

Transcribing and transforming human speech into format useful for computer applications. Currently used in interactive voice response systems and mobile applications. Sample vendors: NICE, Nuance Communications, OpenText, Verint Systems.

3. Virtual Agents

"The current darling of the media," says Forrester, from simple chatbots to advanced systems that can network with humans. Currently used in customer service and support and as a smart home manager. Sample vendors: Amazon, Apple, Artificial Solutions, Assist AI, Creative Virtual, Google, IBM, IPsoft, Microsoft, Satisfi.

4. Machine Learning Platforms

Providing algorithms, APIs, development and training toolkits, data, as well as computing power to design, train, and deploy models into applications, processes, and other machines. Currently used in a wide range of enterprise applications, mostly involving prediction or classification. Sample vendors: Amazon, Fractal Analytics, Google, H2O.ai, Microsoft, SAS, Skytree.

5. AI-optimized Hardware

Graphics processing units（GPU）and appliances specifically designed and architected to efficiently run AI-oriented computational jobs. Currently primarily making a difference in deep learning applications. Sample vendors: Alluviate, Cray, Google, IBM, Intel, Nvidia.

6. Decision Management

Engines that insert rules and logic into AI systems and used for initial setup/training and ongoing maintenance and tuning. A mature technology, it is used in a wide variety of enterprise applications, assisting in or performing automated decision-making. Sample vendors: Advanced Systems Concepts, Informatica, Maana, Pegasystems, UiPath.

7. Deep Learning Platforms

A special type of machine learning consisting of artificial neural networks with multiple abstraction layers. Currently primarily used in pattern recognition and classification applications supported by very large data sets. Sample vendors: Deep Instinct, Ersatz Labs, Fluid AI, MathWorks, Peltarion, Saffron Technology, Sentient Technologies.

8. Biometrics

Enabling more natural interactions between humans and machines, including but not limited to image and touch recognition, speech, and body language. Currently used primarily in market research. Sample vendors: 3VR, Affectiva, Agnitio, FaceFirst, Sensory, Synqera, Tahzoo.

9. Robotic Process Automation

Using scripts and other methods to automate human action to support efficient business processes. Currently used where it's too expensive or inefficient for humans to execute a task or a process. Sample vendors: Advanced Systems Concepts, Automation Anywhere, Blue Prism, UiPath, WorkFusion.

10. Text Analytics and NLP

Natural language processing（NLP）uses and supports text analytics by facilitating the understanding of sentence structure and meaning, sentiment, and intent through statistical and machine learning methods. Currently used in fraud detection and security, a wide range of automated assistants, and applications for mining unstructured data. Sample vendors: Basis Technology, Coveo, Expert System, Indico, Knime, Lexalytics, Linguamatics, Mindbreeze, Sinequa, Stratifyd, Synapsify.

There are certainly many business benefits gained from AI technologies today, but according to a survey Forrester conducted last year, there are also obstacles to AI adoption as expressed by companies with no plans of investing in AI.

There is no defined business case: 42%

Not clear what AI can be used for: 39%

Don't have the required skills: 33%

Need first to invest in modernizing data platform: 29%

Don't have the budget: 23%

Not certain what is needed for implementing an AI system: 19%

AI systems are not proven: 14%

Do not have the right processes or governance: 13%

AI is a lot of hype with little substance: 11%

Don't own or have access to the required data: 8%

Not sure what AI means: 3%

Once enterprises overcome these obstacles, Forrester concludes, they stand to gain from AI driving accelerated transformation in customer-facing applications and developing an interconnected web of enterprise intelligence.

✎ New Words

technology	[tek'nɒlədʒɪ]	n. 科技(总称)；工业技术
flourish	['flʌrɪʃ]	vi. 茂盛，繁荣；活跃，蓬勃
hype	[haɪp]	n. 天花乱坠的广告宣传
		vt. 大肆宣传；夸张地宣传
attention	[ə'tenʃn]	n. 注意，注意力
investment	[ɪn'vestmənt]	n. 投资，投资额
predict	[prɪ'dɪkt]	vt. 预言，预测，预示，预告
estimate	['estɪmɪt]	n. 估计，预测
	['estɪmeɪt]	vt. 估计，估算；评价

sub-discipline	[sʌb-'dɪsəplɪn]	*n*.子学科
time-tested	['taɪm'testɪd]	*adj*.经受时间考验的,久经试验的
decision-making	[dɪ'sɪʒn'meɪkɪŋ]	*n*.决策
		adj.决策的
generation	[ˌdʒenə'reɪʃn]	*n*.产生,生成
insight	['ɪnsaɪt]	*n*.洞察力,洞悉
transcribe	[træn'skraɪb]	*vt*.转录;改编(乐曲)
format	['fɔːmæt]	*n*.格式
		vt.使格式化
interactive	[ˌɪntər'æktɪv]	*adj*.交互式的,互动的;互相作用的,相互影响的
voice	[vɔɪs]	*n*.语音
mobile	[ˌ'məʊbaɪl]	*adj*.可移动的
		n.手机
agent	['eɪdʒənt]	*n*.代理人;代理商
		vt.由……作中介;由……代理
		adj.代理的
toolkit	['tuːlkɪt]	*n*.工具包,工具箱
design	[dɪ'zaɪn]	*v*.&*n*.设计
model	['mɒdl]	*n*.模型;模式;典型
process	['prəʊses]	*n*.过程
classification	[ˌklæsɪfɪ'keɪʃn]	*n*.分类;分级;类别
decision	[dɪ'sɪʒn]	*n*.决定,决策
abstraction	[æb'strækʃn]	*n*.抽象;抽象化;抽象概念
layer	['leɪə]	*n*.层,层次
		vt.把……分层
biometric	[ˌbaɪəʊ'metrɪk]	*n*.计量生物学
recognition	[ˌrekəg'nɪʃn]	*n*.识别,认识
automation	[ˌɔːtə'meɪʃn]	*n*.自动化(技术),自动操作
analytic	[ˌænə'lɪtɪk]	*adj*.分析的,解析的
facilitate	[fə'sɪlɪteɪt]	*vt*.促进,助长
statistical	[stə'tɪstɪkl]	*adj*.统计的,统计学的
obstacle	['ɒbstəkl]	*n*.障碍,障碍物
adoption	[ə'dɒpʃn]	*n*.采用
skill	[skɪl]	*n*.技能,技巧;才能,本领
modernize	['mɒdənaɪz]	*v*.使现代化
budget	['bʌdʒɪt]	*n*.预算;预算拨款
		v.把……编入预算
implement	['ɪmplɪmənt]	*vt*.实施,执行;使生效,实现

		n.工具,手段
prove	[pru:v]	*vt*.证明,证实;显示
		vi.显示出,证明是
governance	['gʌvənəns]	*n*.管理;支配
substance	['sʌbstəns]	*n*.实质,内容
interconnect	[ˌɪntəkə'nekt]	*vi*.互相连接,互相联系
		vt.使互相连接;使互相联系

✎ Phrases

media attention	媒体关注度
coined in	发明于
natural language generation	自然语言生成
mobile application	移动应用
customer service	客户服务
smart home manager	智能管家
as well as	也,又
computing power	计算能力
AI-oriented computational job	面向人工智能的计算任务
consist of...	由……组成
data set	数据集
touch recognition	触摸识别
body language	肢体语言,手势语言
text analytic	文本分析
a wide range of	广泛的
business case	商业案例
a lot of	许多的
customer-facing application	面向客户的应用
enterprise intelligence	企业智能

✎ Abbreviations

IDC（International Data Corporation）	国际数据公司
API（Application Programming Interface）	应用程序编程接口
GPU（Graphics Processing Unit）	图形处理单元,图形处理器
NLP（Natural Language Processing）	自然语言处理

✎ Exercises

【Ex.5】 根据课文内容回答问题。

1. What did a Narrative Science survey find last year?

2. What does artificial intelligence today include?

3. Where is natural language generation currently used?

4. What does speech recognition do? What are the sample vendors?

5. What do machine learning platforms do?

6. Where is decision management used? What are the sample vendors?

7. What are deep learning platforms? Where are they currently primarily used?

8. Where is biotrics currently used? What are the sample vendors?

9. Where is robotic process automation currently use? What are the sample vendors?

10. How does natural language processing use and support text analytics?

Reading

The History of AI

1936：Turing machine

The British mathematician Alan Turing applied his theories to prove that a computing machine—known as a 'Turing machine'—would be capable of executing cognitive① processes，provided they could be broken down into② multiple，individual steps and represented by an algorithm. In doing so，he laid the foundation③ for what we call artificial intelligence today.

1956：The history began：the term 'AI' was coined

In the summer of 1956，scientists gathered for a conference at Dartmouth College in New Hampshire. They believed that aspects of learning as well as other

---

① cognitive ['kɒgnətɪv] *adj*. 认知的

② be broken down into：被分解成

③ foundation [faʊn'deɪʃn] *n*. 基础

characteristics of human intelligence could be simulated by machines. The programmer[1] John McCarthy proposed calling this 'artificial intelligence'. The world's first AI program, 'Logic Theorist[2]', which managed to prove several dozen mathematical theorems and data, was also written during the conference.

1966：Birth of the first chatbot

The German-American computer scientist Joseph Weizenbaum of the Massachusetts Institute of Technology invented a computer program that communicated with humans. 'ELIZA' used scripts to simulate various conversation[3] partners such as a psychotherapist[4]. Weizenbaum was surprised at the simplicity of the means required for ELIZA to create the illusion[5] of a human conversation partner.

1972：AI entered the medical field

With 'MYCIN', artificial intelligence found its way into medical practices: The expert system developed by Ted Shortliffe at Stanford University was used for the treatment of illnesses. Expert systems were computer programs that bundle[6] the knowledge for a specialist field using formulas, rules, and a knowledge database. They were used for diagnosis[7] and treatment support in medicine.

1986：'NETtalk' spoke

The computer was given a voice for the first time. Terrence J. Sejnowski and Charles Rosenberg taught their 'NETtalk' program to speak by inputting sample sentences and phoneme[8] chains. NETtalk was able to read words and pronounce them correctly, and could apply what it had learned to words it did not know. It was one of the early artificial neural networks—programs that were supplied with large datasets and were able to draw their own conclusions on this basis. Their structure and function were thereby similar to those of the human brain.

---

① programmer ['prəʊgræmə] n.程序设计者
② theorist ['θɪərɪst] n.理论家；学说创立人
③ conversation [ˌkɒnvə'seɪʃn] n.交谈，会话
④ psychotherapist [ˌsaɪkəʊ'θerəpɪst] n.心理治疗师
⑤ illusion [ɪ'luːʒn] n.错觉；假象
⑥ bundle ['bʌndl] v.捆绑
⑦ diagnosis [ˌdaɪəg'nəʊsɪs] n.诊断
⑧ phoneme ['fəʊniːm] n.音位，音素

1997：Computer beat world chess champion

The AI chess computer 'Deep Blue' from IBM defeated the incumbent chess world champion Garry Kasparov in a tournament. This was considered a historic success in an area previously dominated by humans. Critics, however, found fault with Deep Blue for winning merely by calculating all possible moves, rather than with cognitive intelligence.

2011：AI entered everyday life

Technology leaps in the hardware and software fields paved the way for artificial intelligence to enter everyday life. Powerful processors① and graphics cards② in computers, smartphones, and tablets gave regular consumers access to AI programs. Digital assistants in particular enjoyed great popularity: Apple's 'Siri' came to the market in 2011, Microsoft introduced the 'Cortana' software in 2014, and Amazon presented Amazon Echo with the voice service 'Alexa' in 2015.

2011：AI 'Watson' won quiz③ show

The computer program 'Watson' competed in a U.S. television quiz show in the form of an animated on-screen symbol and won against the human players. In doing so, Watson proved that it understood natural language and was able to answer difficult questions quickly.

2018：AI debated space travel④ and made a hairdressing appointment

These two examples demonstrated the capabilities of artificial intelligence: In June, 'Project Debater' from IBM debated complex topics with two master debaters⑤—and performed remarkably well. A few weeks before, Google demonstrated at a conference how the AI program 'Duplex' phoned a hairdresser and conversationally made an appointment—without the lady on the other end of the line noticing that she was talking to a machine.

---

① processor ['prəʊsesə] *n.* 处理器
② graphics card：图形卡，显卡
③ quiz [kwɪz] *n.* 智力竞赛
④ space travel：太空旅行
⑤ debater [dɪ'beɪtə] *n.* 辩论者，讨论者

20xx：The near future is intelligent

Decades of research notwithstanding，artificial intelligence is comparatively① still in its infancy. It needs to become more reliable and secure against manipulation before it can be used in sensitive areas，such as autonomous driving or medicine. Another goal is for AI systems to learn to explain their decisions so that humans can comprehend them and better research how AI thinks. Numerous scientists are working on these topics.

参考译文

人 工 智 能

人工智能(AI)是机器、特别是计算机系统对人类智能处理的模拟。这些过程包括学习(获取信息和使用信息的规则)、推理(使用规则来达到近似或明确的结论)和自我校正。人工智能的典型应用包括专家系统、语音识别和机器视觉。

人工智能可以分为弱人工智能与强人工智能两类。弱人工智能，也称为窄人工智能，是为特定任务而设计和训练的人工智能系统。虚拟个人助理，如 Apple 的 Siri，是一种弱人工智能。强人工智能，也称为通用人工智能，是一种具有广泛人类认知能力的人工智能系统。当提出一项不熟悉的任务时，强人工智能系统能够在没有人为干预的情况下找到解决方案。

由于人工智能的硬件、软件和人员成本可能很昂贵，因此许多供应商在其标准产品中包含人工智能组件以及访问人工智能即服务(AIaaS)平台。在做出承诺之前，人工智能即服务允许个人和公司为各种商业目的进行人工智能试验，并对多个平台进行抽样调查。流行的人工智能云产品包括 Amazon 人工智能服务、IBM Watson 助理、Microsoft 认知服务和 Google 人工智能服务。

虽然人工智能工具为企业提供了一系列新功能，但人工智能的使用引发了伦理问题。这是因为深度学习算法是许多最先进的人工智能工具的基础，它们的智能仅仅与训练时所提供的数据匹配。因为由人类选择用何种数据来训练人工智能程序，而人类本身可能有偏见，所以必须密切监控。

一些业内专家认为，人工智能这一术语与流行文化联系太紧密，导致普通大众对人工智能产生不切实际的恐惧，以及对人工智能如何改变工作场所和生活方式抱有不太可能的期望。研究人员和营销人员希望增强智能(具有更中性内涵)会帮助人们明白人工智能

---

① comparatively [kəm'pærətɪvli] *adv*. 相对地，比较地

只能改进产品和服务，而不是取代使用它们的人。

1. 人工智能的类型

密歇根州立大学综合生物学和计算机科学与工程的助理教授 Arend Hintze 将人工智能分为 4 类，从现有的人工智能系统到尚未存在的感觉系统。具体分类如下。

（1）类型 1：反应机器。一个例子是 Deep Blue（深蓝），它是一个在 20 世纪 90 年代击败 Garry Kasparov 的 IBM 国际象棋程序。Deep Blue 可以识别棋盘上的棋子并进行预测，但它没有记忆，也无法使用过去的经验来指导未来的棋子。它分析了自己和对手，并选择最具战略性的举措。Deep Blue 和 Google 的 AlphaGo 专为狭窄目的而设计，不能轻易应用于其他情况。

（2）类型 2：有限的存储。这些人工智能系统可以使用过去的经验来指导未来的决策。自动驾驶汽车的一些决策功能就是这样设计的。观察结果可以告知在不远的将来发生的行动，例如换车道。这些观察结果不会永久存储。

（3）类型 3：心智理论。这个心理学术语指的是他人有自己的信念、欲望和意图，这会影响他们的决策。这种人工智能尚不存在。

（4）类型 4：自我意识。在这个类别中，人工智能系统具有自我意识感，具有知觉。具有自我意识的机器了解其当前状态，并可以使用该信息来推断其他人的感受。这种类型的人工智能尚不存在。

2. 人工智能技术的例子

人工智能被整合到各种不同类型的技术中。这里有一些例子。

（1）自动化：可以使系统或过程自动运行。例如，机器人过程自动化（RPA）可以通过编程来执行人类通常执行的大量可重复的任务。RPA 与 IT 自动化的不同之处在于它可以适应不断变化的环境。

（2）机器学习：使计算机无须编程即可行动的科学。深度学习是机器学习的一个子集，简言之，它可以被认为是自动化进行预测分析。有 3 种类型的机器学习算法：

- 监督学习：标记数据集，以便可以检测模式并用于标记新数据集。
- 无监督学习：不标记数据集，并根据相似性或差异性进行排序。
- 强化学习：不标记数据集，但在执行一个行动或多个行动后，人工智能系统会得到反馈。

（3）机器视觉：让计算机具有视觉的科学。该技术使用相机、模数转换和数字信号处理来捕获和分析视觉信息。它通常被比作人类的视力，但机器视觉不受生物学的约束，例如可以编程以透视墙壁。它用于从签名识别到医学图像分析的各种应用中。计算机视觉是基于机器的图像处理，通常与机器视觉混为一谈。

（4）自然语言处理（NLP）：通过计算机程序处理人类（不是计算机）的语言。其中一个较早且最著名的 NLP 示例是垃圾邮件检测，它查看主题行和电子邮件的文本并确定它是否为垃圾邮件。目前的 NLP 方法基于机器学习。NLP 任务包括文本翻译、情感分析和语音识别。

（5）机器人技术：一个专注于机器人设计和制造的工程领域。机器人通常用于执行人类难以执行或一直执行的任务。它们用于汽车生产的装配线或由 NASA 用于在太空中移动大型物体。研究人员还利用机器学习来构建可在社交场合进行交互的机器人。

（6）自动驾驶汽车：它们把计算机视觉、图像识别和深度学习相结合，使用自动化技能驾驶车辆，遇到意外障碍（例如行人）时在给定车道上停车。

3. 人工智能应用

人工智能已经进入了许多领域。这里列举 6 个示例。

（1）人工智能应用于医疗保健领域。最大的好处是改善患者的治疗效果和降低成本。公司正在应用机器学习来做出比人类更好、更快的诊断。IBM Watson 是最著名的医疗保健技术之一。它理解自然语言，并能够回答所提出的问题。系统挖掘患者数据和其他可用数据源以形成假设，然后它将给出一个置信评分模式。其他人工智能应用程序包括聊天机器人。聊天机器人是一个计算机程序，用于在线回答问题和帮助客户，帮助安排后续预约或自动计费，以及提供基本医疗反馈的虚拟健康助理。

（2）人工智能应用于商业领域。机器人过程自动化正被应用于通常由人类执行的、高度重复的任务。机器学习算法正在集成到分析和客户关系管理平台中，用于发现和分析信息并更好地为客户服务。聊天机器人已被纳入网站，为客户提供即时服务。工作岗位的自动化也成为学术界和 IT 分析师的话题。

（3）人工智能应用于教育领域。人工智能可以自动分级，节省教师时间。人工智能可以评估学生并应对他们的需求，帮助他们按照自己的进度工作。人工智能导师可以为学生提供额外的支持，确保他们处于正确轨道上。人工智能可以改变学生学习的地点和方式，甚至可以取代一些教师。

（4）人工智能应用于金融领域。个人理财应用程序中的人工智能（如 Mint 或 Turbo Tax）正在进入金融机构。这些应用程序收集个人数据并提供财务建议。其他程序（例如 IBM Watson）已经应用于购买房屋的过程。今天，华尔街很大一部分交易都是由软件完成的。

（5）人工智能应用于法律领域。在法律上，对人来说，发现过程（筛选文件）是非常困难的。自动化地完成此项工作可以大大节省时间。创业公司还在构建计算机回答助手，通过编程来检查与数据库相关的分类和本体，筛选出问题的答案。

（6）人工智能应用于制造业。这个领域一直处于将机器人纳入工作流程的最前沿。工业机器人曾经执行单一任务并与人类工作人员分开，但随着技术的进步这一现象已经发生了变化。

4. 安全和伦理问题

人工智能在自动驾驶汽车领域的应用带来了安全和伦理方面的问题。汽车可以被黑客入侵，当自动驾驶汽车涉及事故时，责任也不清楚。自动驾驶车辆也可能处于无法避免事故的情况，迫使编程人员就如何最大限度地减少损坏做出伦理决定。

另一个主要问题是存在滥用人工智能工具的可能性。黑客们开始使用复杂的机器学习工具来访问敏感系统，使安全问题越来越复杂化。

基于深度学习的视频和音频生成工具也为不良行为者提供了所谓换脸所需的工具，他们可以制作公众人物的视频，尽管这些公众人物从未说过这些话，也从未做过这些事，但这些视频却让人不得不信。

5. 人工智能技术的规范

尽管存在这些潜在的风险，但很少有关于使用人工智能工具的法规，而且即便有法规，它们通常也只是间接地涉及人工智能。例如，联邦公平贷款法规要求金融机构向潜在客户解释信用决策，这些法规限制了贷方可以使用深度学习算法的程度，这些算法本质上通常是不透明的。欧洲的 GDPR 严格限制企业使用消费者数据的方法，这阻碍了许多面向消费者的人工智能应用程序的培训和功能。

2016 年，国家科学技术委员会发布了一份报告，研究政府监管在人工智能发展中可能发挥的作用，但并未建议考虑具体立法。从那时起，这个问题就很少受到立法者的关注。

Unit 2

录音

Text A

Popular Search Algorithms

Searching is the universal technique of problem solving in AI. There are some single-player games such as tile games, Sudoku, crossword, etc. The search algorithms help you to search for a particular position in such games.

1. Search Terminology

Problem Space—It is the environment in which the search takes place(A set of states and set of operators to change those states).

Problem Instance—It is initial state + goal state.

Problem Space Graph—It represents problem state. States are shown by nodes and operators are shown by edges.

Depth of a problem—Length of a shortest path or shortest sequence of operators from initial state to goal state.

Space Complexity—The maximum number of nodes that are stored in memory.

Time Complexity—The maximum number of nodes that are created.

Admissibility—A property of an algorithm to always find an optimal solution.

Branching Factor—The average number of child nodes in the problem space graph.

Depth—Length of the shortest path from initial state to goal state.

2. Brute-Force Search Strategies

They are most simple, as they do not need any domain-specific knowledge. They work fine with small number of possible states.

Requirements:

(1) State description;

(2) A set of valid operators;

(3) Initial state;

(4) Goal state description.

2.1　Breadth-First Search

It starts from the root node, explores the neighboring nodes first and moves towards the next level neighbors. It generates one tree at a time until the solution is found. It can be implemented using FIFO queue data structure. This method provides shortest path to the solution.

If branching factor (average number of child nodes for a given node) $= b$ and depth $= d$, then number of nodes at level is b^d.

The total number of nodes created in worst case is $b + b^2 + b^3 + \cdots + b^d$.

Disadvantage: Since each level of nodes is saved for creating the next one, it consumes a lot of memory space. Space requirement to store nodes is exponential.

Its complexity depends on the number of nodes. It can check duplicate nodes.

2.2　Depth-First Search

It is implemented in recursion with LIFO stack data structure. It creates the same set of nodes as breadth-first method, only in the different order.

As the nodes on the single path are stored in each iteration from root to leaf node, the space requirement to store nodes is linear. With branching factor band depth as m, the storage space is bm.

Disadvantage: This algorithm may not terminate and go on infinitely on one path. The solution to this issue is to choose a cut-off depth. If the ideal cut-off is d, and if chosen cut-off is lesser than d, then this algorithm may fail. If chosen cut-off is more than d, then execution time increases.

Its complexity depends on the number of paths. It cannot check duplicate nodes.

2.3　Bidirectional Search

It searches forward from initial state and backward from goal state till both meet to identify a common state.

The path from initial state is concatenated with the inverse path from the goal state. Each search is done only up to half of the total path.

2.4　Uniform Cost Search

Sorting is done in increasing cost of the path to a node. It always expands the least cost node. It is identical to breadth-first search if each transition has the same cost.

It explores paths in the increasing order of cost.

Disadvantage：There can be multiple long paths. Uniform cost search must explore them all.

2.5　Iterative Deepening Depth-First Search

It performs depth-first search to level 1，starts over，executes a complete depth-first search to level 2，and continues in such way till the solution is found.

It never creates a node until all lower nodes are generated. It only saves a stack of nodes. The algorithm ends when it finds a solution at depth d. The number of nodes created at depth d is b^d and at depth $d-1$ is b^{d-1}.

2.6　Comparison of Various Algorithms Complexities

Let us see the performance of algorithms based on various criteria(See Table 2-1).

Table 2-1　Performance of Algorithms

Criterion	Breadth First	Depth First	Bidirectional	Uniform Cost	Interactive Deepening
Time	b^d	b^m	$b^{d/2}$	b^d	b^d
Space	b^d	b^m	$b^{d/2}$	b^d	b^d
Optimality	Yes	No	Yes	Yes	Yes
Completeness	Yes	No	Yes	Yes	Yes

3. Informed（Heuristic）Search Strategies

To solve large problems with large number of possible states，problem-specific knowledge needs to be added to increase the efficiency of search algorithms.

3.1　Heuristic Evaluation Functions

They calculate the cost of optimal path between two states. A heuristic function for sliding-tiles games is computed by counting the number of moves that each tile makes from its goal state and adding the number of moves for all tiles.

3.2 Pure Heuristic Search

It expands nodes in the order of their heuristic values. It creates two lists, a closed list for the already expanded nodes and an open list for the created but unexpanded nodes.

In each iteration, a node with a minimum heuristic value is expanded, all its child nodes are created and placed in the closed list. Then, the heuristic function is applied to the child nodes and they are placed in the open list according to their heuristic value. The shorter paths are saved and the longer ones are disposed.

3.3 A * Search

It is the best-known form of best-first search. It avoids expanding paths that are already expensive, but expands most promising paths first.

$f(n) = g(n) + h(n)$, where

(1) $g(n)$ is the cost (so far) to reach the node.

(2) $h(n)$ is estimated cost to get from the node to the goal.

(3) $f(n)$ is estimated total cost of path through n to goal. It is implemented using priority queue by increasing $f(n)$.

3.4 Greedy Best First Search

It expands the node that is estimated to be closest to goal. It expands nodes based on $f(n) = h(n)$. It is implemented using priority queue.

Disadvantage: It can get stuck in loops. It is not optimal.

4. Local Search Algorithms

They start from a prospective solution and then move to a neighboring solution. They can return a valid solution even if it is interrupted at any time before they end.

4.1 Hill-Climbing Search

It is an iterative algorithm that starts with an arbitrary solution to a problem and attempts to find a better solution by changing a single element of the solution incrementally. If the change produces a better solution, an incremental change is taken as a new solution. This process is repeated until there are no further improvements.

Function Hill-Climbing（problem）returns a state that is a local maximum.

```
inputs: problem, a problem
local variables: current, a node
                 neighbor, a node
current < - Make_Node(Initial - State[problem])
loop
    do neighbor < - a highest_valued successor of current
        if Value[neighbor] ≤ Value[current] then
        return State[current]
        current < - neighbor

end
```

Disadvantage：This algorithm is neither complete nor optimal.

4.2　Local Beam Search

In this algorithm，it holds k number of states at any given time. At the start，these states are generated randomly. The successors of these k states are computed with the help of objective function. If any of these successors is the maximum value of the objective function，then the algorithm stops.

Otherwise the states（initial k states and k number of successors of the states = 2k）are placed in a pool. The pool is then sorted numerically. The highest k states are selected as new initial states. This process continues until a maximum value is reached.

Function BeamSearch（problem，k）returns a solution state.

```
start with k randomly generated states
    loop
    generate all successors of all k states
        if any of the states = solution, then return the state
    else select the k best successors
end
```

4.3　Simulated Annealing

Annealing is the process of heating and cooling a metal to change its internal structure for modifying its physical properties. When the metal cools，its new structure is seized，and the metal retains its newly obtained properties. In simulated annealing process，the temperature is kept variable.

We initially set the temperature high and then allow it to 'cool' slowly as the algorithm proceeds. When the temperature is high，the algorithm is allowed to accept worse solutions with high frequency.

```
start
    initialize k = 0; L = integer number of variables;
    from i→j, search the performance difference Δ;
    if Δ <= 0 then accept else if exp(-Δ/T(k))>random(0,1) then accept;
    repeat steps 1 and 2 for L(k) steps;
    k = k + 1.
    repeat steps 1 through 4 till the criteria is met.
end
```

4.4 Travelling Salesman Problem

In this algorithm, the objective is to find a low-cost tour that starts from a city, visits all cities en-route exactly once and ends at the same starting city.

```
start
    find out all (n - 1)! Possible solutions, where n is the total number of cities.
    determine the minimum cost by finding out the cost of each of these (n - 1)! solutions.
    finally, keep the one with the minimum cost.
end
```

✐ New Words

search	[sɜːtʃ]	v. 搜索，搜寻；调查
		n. 搜索
algorithm	[ˈælgərɪðəm]	n. 算法
universal	[ˌjuːnɪˈvɜːsl]	adj. 普遍的，一般的，通用的
crossword	[ˈkrɒswɜːd]	n. 填字游戏，纵横字谜
position	[pəˈzɪʃn]	n. 位置，方位；态度；状态
		vt. 安置；把…放在适当位置；给…定位
state	[steɪt]	n. 状态
operator	[ˈɒpəreɪtə]	n. 运算符
represent	[ˌreprɪˈzent]	vt. 代表，表现
node	[nəʊd]	n. 节点
edge	[edʒ]	n. 边
depth	[depθ]	n. 深度
sequence	[ˈsiːkwəns]	n. 序列；顺序；连续
		vt. 使按顺序排列
complexity	[kəmˈpleksətɪ]	n. 复杂度，复杂性
admissibility	[ədˌmɪsəˈbɪlətɪ]	n. 可容许性，可接受性
graph	[grɑːf]	n. 图表，曲线图
		vt. 用曲线图表示

factor	['fæktə]	n. 因素,因子
description	[dɪ'skrɪpʃn]	n. 描述,形容
valid	['vælɪd]	adj. 有效的
neighboring	['neɪbərɪŋ]	adj. 邻近的
solution	[sə'luːʃn]	n. 解决;答案
queue	[kjuː]	n. 队列
disadvantage	[ˌdɪsəd'vɑːntɪdʒ]	n. 缺点,劣势,短处
consume	[kən'sjuːm]	vt. 消耗,消费
exponential	[ˌekspə'nenʃl]	adj. 指数的,幂数的;越来越快的
		n. 指数
check	[tʃek]	vt. 检查,核对
duplicate	['djuːplɪkeɪt]	v. 重复
recursion	[rɪ'kɜːʃn]	n. 递归,递推
stack	['stæk]	n. 堆栈
linear	['lɪnɪə]	adj. 线性的
terminate	['tɜːmɪneɪt]	v. 结束,使终结
infinitely	['ɪnfɪnətlɪ]	adv. 无限地,无穷地
cut-off	['kʌtɔf]	n. 截止;界限
bidirectional	[ˌbaɪdɪ'rekʃənl]	adj. 双向的
inverse	[ˌɪn'vɜːs]	adj. 相反的;逆向的;倒转的
		n. 相反;倒转;相反的事物
		vt. 使倒转
iterative	['ɪtərətɪv]	adj. 迭代的,重复的,反复的
		n. 反复体
optimality	[ɒptɪ'mælɪtɪ]	n. 最优性;最佳性
inform	[ɪn'fɔːm]	vt. 通知
heuristic	[hjʊ''rɪstɪk]	adj. 启发式的;探试的,探索的
evaluation	[ɪˌvæljʊ'eɪʃn]	n. 评估,估价
list	[lɪst]	n. 列表;清单,目录
		vt. 列出
unexpanded	[ʌnɪk'spændɪd]	adj. 未被扩大的,未展开的
dispose	[dɪ'spəʊz]	v. 处理,处置;安排
estimate	['estɪmeɪt]	vt. 估计,估算;评价
prospective	[prə'spektɪv]	adj. 预期的;可能的;有希望的
interrupt	[ˌɪntə'rʌpt]	v.&n. 中断;暂停
arbitrary	['ɑːbɪtrərɪ]	adj. 随意的,任性的
incrementally	[ˌɪnkrɪ'mentəlɪ]	adv. 逐渐地
repeat	[rɪ'piːt]	v. 重复

improvement	[ɪm'pruːvmənt]	n.改进,改善,改良,增进
successor	[sək'sesə]	n.接替的人或事物;继任者
pool	[puːl]	n.水池
numerically	[njuː'merɪklɪ]	adv.用数表示地,数字化地
return	[rɪ'tɜːn]	v.返回,回来;退还;重现
anneal	[ə'niːl]	n.退火
		vt.使退火
internal	[ɪn'tɜːnl]	adj.内部的
frequency	['friːkwənsɪ]	n.频率,次数;频率分布

✏ Phrases

tile game	智力拼图
search algorithm	搜索算法
problem space	问题空间
initial state	起始状态,初态
goal state	目标状态
shortest path	最短路径
space complexity	空间复杂度
time complexity	时间复杂度
branching factor	分支因子,分支系数
brute-force search	蛮力搜索,强力搜索
breadth-first search	宽度优先搜索
root node	根节点
worst case	最坏情况,最坏条件
depth-first search	深度优先搜索
data structure	数据结构
leaf node	叶节点
bidirectional search	双向搜索,双向查找
be concatenated with …	与……连接
uniform cost search	等代价搜索,一致代价搜索
iterative deepening depth-first search	迭代深化深度优先搜索
informed (heuristic) search	启发式搜索
sliding-tiles game	滑动拼图游戏
pure heuristic search	纯启发式搜索
child node	子节点
according to	根据,按照
greedy best first search	贪婪最佳优先搜索

priority queue	优先队列
local search algorithm	局部搜索算法
hill-climbing search	爬山算法
local beam search	局部集束搜索
simulated annealing	模拟退火
physical property	物理属性
travelling salesman problem	旅行商问题

✎ Abbreviations

FIFO（First Input First Output）	先入先出
LIFO（Last In First Out）	后进先出

✎ Exercises

【Ex. 1】 根据课文内容回答问题。

1. What is problem space?

2. What are the requirements for brute-force search strategies?

3. What is the disadvantage of breadth-first search?

4. How does depth-first search work?

5. What is the disadvantage of uniform cost search?

6. How is a heuristic function for sliding-tiles games computed?

7. What is A* search?

8. What is hill-climbing search?

9. What is annealing?

10. What is travelling salesman problem?

【Ex. 2】 把下列单词或词组中英互译。

1. admissibility	1. _____
2. algorithm	2. _____
3. complexity	3. _____
4. exponential	4. _____
5. interrupt	5. _____
6. n.最优性；最佳性	6. _____
7. n.因素,因子	7. _____
8. n.递归,递推	8. _____
9. 贪婪最佳优先搜索	9. _____
10. 局部集束搜索	10. _____

【Ex.3】 短文翻译。

Artificial Intelligence

The first thing we need to do is to understand what an AI actually is. The term 'artificial intelligence' refers to a specific field of computer engineering that focuses on creating systems capable of gathering data and making decisions and/or solving problems. An example of basic AI is a computer that can take 1000 photos of cats for input, determine what makes them similar, and then find photos of cats on the internet. The computer has learned, as best as it can, what a photo of a cat looks like and uses this new intelligence to find things that are similar.

Deep learning is what happens when a neural network gets to work. As the layers process data the AI gains a basic understanding. You might be teaching your AI to understand cats, but once it learns what paws are that AI can apply that knowledge to a different task. Deep learning means that instead of understanding what something is, the AI begins to learn 'why'.

【Ex.4】 将下列词填入适当的位置(每词只用一次)。

resources	different	science	considered	operations
development	problem	traveling	huge	referred

Traveling Salesman Problem

The traveling salesman problem is a traditional issue that has something to do with making the most efficient use of resources while at the same time expending the least amount of energy in that utilization. The designation for this type of problem hails back to the days of the __1__ salesman, who often wished to arrange travel in a manner that allowed for visiting the most towns without having to double back and cross into any given town more than once.

In a wider sense, the traveling salesman problem is __2__ to be a classic example of what is known as a tour problem. Essentially, any type of tour problem involves making a series of stops along a designated route and making a return journey without ever making a second visit to any previous stop. Generally, a tour problem is present when there is concern on making the most of available __3__ such as time and mode of travel to accomplish the most in results. Finding a solution to a tour problem is sometimes __4__ to as discovering the least-cost path, implying that the strategic planning of the route will ensure maximum benefit with minimum expenditure incurred.

The concept of the traveling salesman problem can be translated into a number of

　　 5 　disciplines. For example，the idea of combinatorial optimization has a direct relationship to the traveling salesman model. As a form of optimization that is useful in both mathematical and computer 　6 　disciplines，combinatorial optimization seeks to team relevant factors and apply them in a manner that will yield the best results with repeated usage.

　　In a similar manner，discrete optimization attempts to accomplish the same goal，although the term is sometimes employed to refer to tasks or 　7 　that occur on a one-time basis rather than recurring. Discrete optimization also is helpful in computer science and mathematical disciplines. In addition，discrete optimization has a direct relationship to computational complexity theory and is understood to be of use in the 　8 　of artificial intelligence.

　　While the imagery associated with a traveling salesman 　9 　may seem an oversimplification of these types of detailed options for optimization，the idea behind the imagery helps to explain a basic fundamental to any type of optimization that strives for efficiency. The traveling salesman problem that is solved will yield 　10 　benefits in the way of maximum return for minimum investment of resources.

Text B

Pros and Cons of Artificial Intelligence

　　Artificial intelligence can help alleviate the difficulties faced by man but intelligent machines can never be human.

　　Intelligence is best defined as the ability of an individual to adapt his/her behavior to new circumstances. Human intelligence is not a single ability but is rather a composition of abilities like learning, reasoning, problem solving, perception, and understanding of language.

1. Pros

　　(1) With artificial intelligence，the chances of error are almost nil and greater precision and accuracy is achieved.

　　(2) Artificial intelligence finds applications in space exploration. Intelligent robots can be used to explore space. They are machines and hence have the ability to endure the hostile environment of the interplanetary space. They can be made to adapt in such a way that planetary atmospheres do not affect their physical state and functioning.

(3) Intelligent robots can be programmed to reach the Earth's nadirs. They can be used to dig for fuels. They can be used for mining purposes. The intelligence of machines can be harnessed for exploring the depths of oceans. These machines can be of use in overcoming the limitations that humans have.

(4) Intelligent machines can replace human beings in many areas of work. Robots can do certain laborious tasks. Painstaking activities, which have long been carried out by humans, can be taken over by the robots. Owing to the intelligence programmed in them, the machines can shoulder greater responsibilities and can be programmed to manage themselves.

(5) Smartphones are a great example of the application of artificial intelligence. In utilities like predicting what a user is going to type and correct human errors in spelling, machine intelligence is at work. Applications like Siri that act as personal assistants, GPS and Maps applications that give users the best or the shortest routes to take as well as the traffic and time estimates to reach there use artificial intelligence. Applications on phones or computers that predict user actions and also make recommendations that suit user choice are applications of AI. Thus, we see that artificial intelligence has made daily life a lot easier.

(6) Fraud detection in smart card-based systems is possible with the use of AI. It is also employed by financial institutions and banks to organize and manage records.

(7) Organizations use avatars that are digital assistants who interact with the users, thus saving the need of human resources.

(8) Emotions that often intercept rational thinking of a human being are not a hindrance for artificial thinkers. Lacking the emotional side, robots can think logically and take the right decisions. Sentiments are associated with moods that affect human efficiency. This is not the case with machines with artificial intelligence.

(9) Artificial intelligence can be utilized in carrying out repetitive and time-consuming tasks efficiently.

(10) Intelligent machines can be employed to do certain dangerous tasks. They can adjust their parameters such as their speed and time, and be made to act quickly, unaffected by factors that affect humans.

(11) When we play a computer game or operate a computer-controlled bot, we are in fact interacting with artificial intelligence. In a game where the computer plays as our opponent, it is with the help of AI that the machine plans the game moves in response to ours. Thus, gaming is among the most common examples of the advantages of artificial intelligence.

(12) AI is at work in the medical field too. Algorithms can help the doctors assess patients and their health risks. It can help them know the side effects that various medicines can have. Surgery simulators use machine intelligence in training medical

professionals. AI can be used to simulate brain functioning, and thus prove useful in the diagnosis and treatment of neurological problems. As in case of any other field, repetitive or time-consuming tasks can be managed through the application of artificial intelligence.

(13) Robotic pets can help patients with depression and also keep them active.

(14) Robotic radio surgery helps achieve precision in the radiation given to tumors, thus reducing the damage to surrounding tissues.

(15) The greatest advantage of artificial intelligence is that machines do not require sleep or breaks, and are able to function without stopping. They can continuously perform the same task without getting bored or tired. When employed to carry out dangerous tasks, the risk to human health and safety is reduced.

2. Cons

(1) One of the main disadvantages of artificial intelligence is the cost incurred in the maintenance and repair. Programs need to be updated to suit the changing requirements, and machines need to be made smarter. In case of a breakdown, the cost of repair may be very high. Procedures to restore lost code or data may be time-consuming and costly.

(2) An important concern regarding the application of artificial intelligence is about ethics and moral values. Is it ethically correct to create replicas of human beings? Do our moral values allow us to recreate intelligence? Intelligence is a gift of nature. It may not be right to install it into a machine to make it work for our benefit.

(3) Machines may be able to store enormous amounts of data, but the storage, access, and retrieval is not as effective as in case of the human brain. They may be able to perform repetitive tasks for long, but they do not get better with experience, like humans do. They are not able to act anything different from what they are programmed to do. Though this is mostly seen as an advantage, it may work the other way, when a situation demands one to act in way different from the usual. Machines may not be as efficient as humans in altering their responses depending on the changing situations.

(4) The idea of machines replacing human beings sounds wonderful. It appears to save us from all the pain. But is it really so exciting? Ideas like working wholeheartedly, with a sense of belonging, and with dedication have no existence in the world of artificial intelligence. Imagine robots working in hospitals. Do you picture them showing the care and concern that humans would? Do you think online assistants (avatars) can give the kind of service that a human being would? Concepts such as care, understanding, and togetherness cannot be understood by machines,

which is why how much ever intelligent they become, they will always lack the human touch.

(5) Imagine intelligent machines employed in creative fields. Do you think robots can excel or even compete with the human mind in creative thinking or originality? Thinking machines lack a creative mind. Human beings are emotional intellectuals. They think and feel. Their feelings guide their thoughts. This is not the case with machines. The intuitive abilities that humans possess, the way humans can judge based on previous knowledge, and the inherent abilities that they have cannot be replicated by machines. Also, machines lack common sense.

(6) If robots begin to replace humans in every field, it will eventually lead to unemployment. People will be left with nothing to do. So much empty time may result in its destructive use. Thinking machines will govern all the fields and populate the positions that humans occupy, leaving thousands of people jobless.

(7) Also, due to the reduced need to use their intelligence, lateral thinking and multitasking abilities of humans may diminish. With so much assistance from machines, if humans do not need to use their thinking abilities, these abilities will gradually decline. With the heavy application of artificial intelligence, humans may become overly dependent on machines, losing their mental capacities.

(8) If the control of machines goes in the wrong hands, it may cause destruction. Machines won't think before acting. Thus, they may be programmed to do the wrong things, or for mass destruction.

(9) Apart from all these cons of AI, there is a fear of robots superseding humans. Ideally, human beings should continue to be the masters of machines. However, if things turn the other way round, the world will turn into chaos. Intelligent machines may prove to be smarter than us people, they might enslave us and start ruling the world.

It should be understood that artificial intelligence has several pros but it has its disadvantages as well. Its benefits and risks should be carefully weighed before employing it for human convenience.

✍ New Words

alleviate	[ə'li:vɪeɪt]	vt. 减轻,缓和
individual	[ˌɪndɪ'vɪdʒuəl]	adj. 个人的；独特的；个别的
		n. 个人；个体
behavior	[bɪ'heɪvjə]	n. 行为；态度
composition	[ˌkɒmpə'zɪʃn]	n. 组合方式；成分；构成
precision	[prɪ'sɪʒn]	n. 精确度,准确(性)

		adj.精确的,准确的
accuracy	['ækjərəsɪ]	n.精确(性),准确(性)
robot	['rəʊbɒt]	n.机器人;遥控装置;自动机
endure	[ɪn'djʊə]	vt.忍耐;容忍
planetary	['plænətrɪ]	adj.行星的
atmosphere	['ætməsfɪə]	n.大气,空气;大气层
nadir	['neɪdɪə]	n.最低点
dig	[dɪg]	vt.挖掘;发掘
fuel	['fjuːəl]	n.燃料
harness	['hɑːnɪs]	vt.利用;控制
overcome	[ˌəʊvə'kʌm]	v.战胜,克服;压倒
laborious	[lə'bɔːrɪəs]	adj.费力的;勤劳的;辛苦的
painstaking	['peɪnzteɪkɪŋ]	adj.辛苦的;苦干的
		n.苦干,刻苦;勤勉;辛苦
utility	[juː'tɪlətɪ]	n.功用,效用
route	[ruːt]	n.路径,途径
action	['ækʃn]	n.行为,行动;功能,作用
recommendation	[ˌrekəmen'deɪʃn]	n.推荐;建议
suit	[suːt]	vt.适合于
avatar	['ævətɑː]	n.化身
emotion	[ɪ'məʊʃn]	n.情感,情绪
intercept	[ˌɪntə'sept]	vt.拦截,拦阻
hindrance	['hɪndrəns]	n.妨害,障碍;障碍物
sentiment	['sentɪmənt]	n.感情,情绪;意见,观点
dangerous	['deɪndʒərəs]	adj.危险的
adjust	[ə'dʒʌst]	v.调整,校正
unaffected	[ˌʌnə'fektɪd]	adj.不受影响的
diagnosis	[ˌdaɪəg'nəʊsɪs]	n.诊断;判断;结论
depression	[dɪ'preʃn]	n.萎靡不振,沮丧
bored	[bɔːd]	adj.无聊的,无趣的,烦人的
incur	[ɪn'kɜː]	vt.引起,招致
repair	[rɪ'peə]	vt.修理;恢复
		n.修理;修理工作;维修状态
breakdown	['breɪkdaʊn]	n.损坏,故障,崩溃,倒塌
moral	['mɒrəl]	adj.道德的
recreate	[ˌriːkrɪ'eɪt]	v.重现;重建;再创造
benefit	['benɪfɪt]	n.利益,好处
		vt.有益于,有助于;使受益;得益

enormous	[ɪ'nɔːməs]	adj.巨大的，庞大的
experience	[ɪk'spɪərɪəns]	n.体验，经验；经历，阅历
excite	[ɪk'saɪt]	vt.使兴奋；激发；刺激；使紧张不安
wholeheartedly	[ˌhəʊl'hɑːtɪdlɪ]	adv.全心全意地，全神贯注地；真心诚意
dedication	[ˌdedɪ'keɪʃn]	n.奉献，献身精神
togetherness	[tə'geðənəs]	n.亲密无间；和睦；团结
originality	[əˌrɪdʒə'næləti]	n.独创性，创造性
emotional	[ɪ'məʊʃənl]	adj.表现强烈情感的；令人动情的；易动感情的
intellectual	[ˌɪntə'lektʃuəl]	adj.智力的，有才智的
		n.知识分子；脑力劳动者
judge	[dʒʌdʒ]	v.评判；断定
inherent	[ɪn'hɪərənt]	adj.固有的，内在的；天生
unemployment	[ˌʌnɪm'plɔɪmənt]	n.失业；失业率；失业状况
jobless	['dʒɒbləs]	adj.没有工作的，失业的
diminish	[dɪ'mɪnɪʃ]	vt.(使)减少，缩小
		vi.变小或减少
decline	[dɪ'klaɪn]	n.衰退；下降
mental	['mentl]	adj.精神的，思想的，心理的；智慧的
destruction	[dɪ'strʌkʃn]	n.破坏，毁灭，消灭，灭亡
supersede	[ˌsuːpə'siːd]	vt.取代，接替
chaos	['keɪɒs]	n.混乱，紊乱；一团糟
enslave	[ɪn'sleɪv]	vt.奴役；征服
ruling	['ruːlɪŋ]	adj.统治的；支配的；管辖的
		n.统治；支配
weigh	[weɪ]	v.权衡，考虑

✎Phrases

problem solving	解决问题
space exploration	空间探索
hostile environment	有害环境
interplanetary space	太空
digital assistant	数字助理
interact with ...	与……相互作用，与……相互影响，与……相互配合
human resource	人力资源
rational thinking	理性思维
time-consuming task	耗时任务

computer-controlled bot　　　　计算机控制的机器人
surrounding tissue　　　　　　外围组织,周围组织

Abbreviations

GPS（Global Positioning System）　全球定位系统

Exercises

【Ex.5】 根据课文内容填空。

1. Artificial intelligence can help alleviate _____ faced by man but _____ can never be human.

2. With artificial intelligence，the chances of error are _____ and greater precision and _____ is achieved.

3. Robots can do certain _____. Painstaking activities, which have long been carried out by humans, can be taken over by _____.

4. Applications on phones or computers that predict _____ and also make recommendations that suit _____ are applications of AI. Thus，we see that artificial intelligence has made daily life _____.

5. Algorithms can help the doctors _____ and _____. It can help them know the _____ that various medicines can have. Surgery simulators use machine intelligence in training _____. AI can be used to simulate _____, and thus prove useful in the diagnosis and treatment of neurological problems.

6. The greatest advantage of artificial intelligence is that machines _____, and are able to function without stopping. They can continuously _____ without getting bored or tired. When employed to carry out dangerous tasks, the risk to _____ is reduced.

7. One of the main disadvantages of artificial intelligence is _____.

8. An important concern regarding the application of artificial intelligence is about _____.

9. If robots begin to replace humans in every field，it will eventually lead to _____. People will be left with _____ to do. So much empty time may result in _____.

10. If the control of machines goes in the wrong hands，it may cause _____. Machines won't _____. Thus, they may be programmed to do _____, or for mass destruction.

Reading

Artificial Intelligence Terms

1. Artificial intelligence

The first thing we need to do is understand what an AI actually is. The term 'artificial intelligence' refers to a specific field of computer engineering[①] that focuses on creating systems capable of gathering data[②] and making decisions and/or solving problems. An example of basic AI is a computer that can take 1000 photos of cats for input, determine what makes them similar, and then find photos of cats on the internet. The computer has learned, as best as it can, what a photo of a cat looks like and uses this new intelligence to find things that are similar.

2. Autonomy

Simply put, autonomy[③] means that an AI construct doesn't need help from people. Driverless cars illustrate the term 'autonomy' in various degrees. Level Four autonomy represents a vehicle that doesn't need a steering wheel[④] or pedals: it doesn't need a human inside of it to operate at full capacity. If we ever have a vehicle that can operate without a driver, and also doesn't need to connect to any grid, server, GPS, or other external source in order to function it'll have reached Level Five autonomy.

3. Algorithm

The most important part of AI is the algorithm. These are math formulas[⑤] and/or programming commands that inform[⑥] a regular non-intelligent computer on how to solve problems with artificial intelligence. Algorithms are rules that teach computers how to figure things out on their own. It may be a nerdy construct of numbers and

---

① computer engineering：计算机工程
② gathering data：收集数据
③ autonomy [ɔː'tɒnəmɪ] n.自治
④ steering wheel：方向盘
⑤ math formulas：数学公式
⑥ inform [ɪn'fɔːm] vt.通知

commands.

4．Machine learning

The meat and potatoes of AI is machine learning—in fact it's typically acceptable to substitute the terms artificial intelligence and machine learning for one another. They aren't quite the same, however, but connected.

Machine learning is the process by which an AI uses algorithms to perform artificial intelligence functions. It's the result of applying rules[①] to create outcomes through an AI.

5．Black box

When the rules are applied an AI does a lot of complex math. This math, often, can't even be understood by humans (and sometimes it just wouldn't be worth the time it would take for us to figure it all out) yet the system outputs useful information. When this happens it's called black box learning. The real work happens in such a way that we don't really care how the computer arrived at the decisions it's made, because we know what rules it used to get there.

6．Neural network

When we want an AI to get better at something we create a neural network. These networks are designed to be very similar to the human nervous system and brain. It uses stages of learning to give AI the ability to solve complex problems by breaking them down into levels of data. The first level of the network may only worry about a few pixels in an image file[②] and check for similarities in other files. Once the initial stage is done, the neural network will pass its findings to the next level which will try to understand a few more pixels, and perhaps some metadata[③]. This process continues at every level of a neural network.

7．Deep learning

Deep learning is what happens when a neural network gets to work. As the layers

---

① rule [ruːl] n. 规则

② image file：图像文件

③ metadata ['metədeɪtə] n. 元数据

process data the AI gains a basic understanding. You might be teaching your AI to understand cats, but once it learns what paws are that AI can apply that knowledge to a different task. Deep learning means that instead of understanding what something is, the AI begins to learn 'why'.

8. Natural language processing

It takes an advanced neural network to parse human language. When an AI is trained to interpret human communication, it's called natural language processing. This is useful for chat bots and translation services[①], but it's also represented at the cutting edge[②] by AI assistants like Alexa and Siri.

9. Reinforcement learning

AI is a lot more like humans than we might be comfortably believing. We learn in almost the exact same way. One method of teaching a machine, just like a person, is to use reinforcement learning[③]. This involves giving the AI a goal that isn't defined with a specific metric, such as telling it to 'improve efficiency' or 'find solutions.' Instead of finding one specific answer the AI will run scenarios and report results, which are then evaluated by humans and judged[④]. The AI takes the feedback and adjusts the next scenario to achieve better results.

10. Supervised learning

This is the very serious business of proving things. When you train an AI model using a supervised learning method you provide the machine with the correct answer ahead of time. Basically the AI knows the answer and it knows the question. This is the most common method of training because it yields the most data: it defines patterns between the question and answer.

If you want to know why something happens, or how something happens, an AI can look at the data and determine connections using the supervised learning method.

---

① translation service：翻译服务
② cutting edge：尖端，最前沿
③ reinforcement learning：强化学习
④ judge [dʒʌdʒ] v. 评判；断定

11. Unsupervised learning

In many ways the spookiest[①] part of AI research is realizing that the machines are really capable of learning, and they're using layers upon layers of data and processing capability to do so. With unsupervised learning we don't give the AI an answer. Rather than finding patterns that are predefined[②] like, "why people choose one brand over another," we simply feed a machine a bunch of data so that it can find whatever patterns it is able to.

参考译文

热门搜索算法

搜索是人工智能中解决问题的通用技术。有一些单人游戏,如智力拼图、数独游戏、填字游戏等。搜索算法可以帮助你搜索此类游戏中的特定位置。

1. 搜索术语

问题空间——这是搜索发生的环境(一组状态和一组运算符来改变这些状态)。

问题实例——它是初始状态 + 目标状态。

问题空间图——它代表问题状态。状态由节点显示,运算符由边显示。

问题的深度——从初始状态到目标状态的最短路径或最短操作符序列的长度。

空间复杂度——存储在内存中的最大节点数。

时间复杂度——创建的最大节点数。

可容许性——总是找到最优解的算法的属性。

分支因子——问题空间图中的平均子节点数。

深度——从初始状态到目标状态的最短路径的长度。

2. 蛮力搜索策略

它们很简单,因为它们不需要任何特定领域的知识。在可能状态很少时,这些策略颇为适用。

要求:

---

① spooky ['spuːki] *adj*. 不可思议的

② predefined [priːdɪ'faɪnd] *n*. 预定义

（1）状态描述；

（2）一组有效的运算符；

（3）初始状态；

（4）目标状态描述。

2.1　广度优先搜索

它从根节点开始，首先探索相邻节点并向下一级邻居移动。它一次生成一棵树，直到找到解决方案。它可以使用 FIFO 队列数据结构实现。该方法提供了解决问题的最短路径。

如果分支因子（给定节点的子节点的平均数量）$= b$ 且深度 $= d$，则该级别的节点数是 b^d。

在最坏情况下创建的节点总数是 $b + b^2 + b^3 + \cdots + b^d$。

缺点：由于保存了每个级别的节点以创建下一个节点，因此会消耗大量内存空间。存储节点的空间要求是指数级的。

其复杂性取决于节点数量。它可以检查重复的节点。

2.2　深度优先搜索

它用 LIFO 堆栈数据结构以递归方式实现。它创建与广度优先方法相同的节点集，只是顺序不同。

由于单个路径上的节点存储在从根节点到叶节点的每次迭代中，因此存储节点的空间要求是线性的。当分支因子带深度为 m 时，存储空间为 bm。

缺点：此算法可能无法终止并在一条路径上无限循环。解决这一问题的方法是选择截止深度。如果理想截止值为 d 且当选择的截止值小于 d，则该算法可能失败。当选择的截止值大于 d，则会增加执行时间。

其复杂性取决于路径的数量。它无法检查重复的节点。

2.3　双向搜索

它从初始状态向前搜索，从目标状态向后搜索，直到两者相遇以识别共同状态。

从初始状态开始的路径与来自目标状态的反向路径相关联。每次搜索仅完成总路径的一半。

2.4　等代价搜索

排序是通过增加节点路径的代价来完成的。它总是扩展最低代价节点。如果每个转换都具有相同的代价，则与广度优先搜索相同。

它以代价增加的顺序探索路径。

缺点：可能有多个长路径。等代价搜索必须全部探索。

2.5　迭代深化深度优先搜索

它执行深度优先搜索到级别 1，再重新开始，执行完整的深度优先搜索到级别 2，并继续以这种方式直到找到解决方案。

　　直到生成所有较低节点,它才会创建一个节点。它只保存节点堆栈。算法在深度 d 处找到解时结束。在深度 d 处创建的节点的数量是 b^d,且在深度 $d-1$ 处是 b^{d-1}。

2.6　各种算法复杂性的比较

让我们看看基于各种标准的算法性能。

标准	广度优先	深度优先	双向	等代价	迭代深化
时间	b^d	b^m	$b^{d/2}$	b^d	b^d
空间	b^d	b^m	$b^{d/2}$	b^d	b^d
最优性	Yes	No	Yes	Yes	Yes
完整性	Yes	No	Yes	Yes	Yes

3. 知情(启发式)搜索策略

为了解决具有大量可能状态的大问题,需要添加针对问题的知识以提高搜索算法的效率。

3.1　启发式评估函数

其计算两个状态之间最佳路径的代价。滑动拼图游戏的启发式函数由计算移动滑块的数量来完成,每个滑块都来自其目标状态,还要加上全部滑块的移动数。

3.2　纯启发式搜索

它按照启发式值的顺序扩展节点。它创建了两个列表,一个是已经展开的节点的闭合列表,另一个是已创建但未展开的节点的开放列表。

在每次迭代中,扩展具有最小启发式值的节点,创建其所有子节点并将其放置在闭合列表中。然后,将启发式函数应用于子节点,并根据它们的启发式值将它们放置在打开列表中。保存较短的路径,并处理较长的路径。

3.3　A＊搜索

它是最佳优先搜索的最知名形式。它避免扩展那些很昂贵的路径,而是首先扩展最有希望的路径。

$f(n)=g(n)+h(n)$,其中:

(1) $g(n)$ 是到达节点的成本(到目前为止)。

(2) $h(n)$ 是从节点到目标的估计成本。

(3) $f(n)$ 是估计从 n 到目标的路径总成本。它通过增加 $f(n)$ 使用优先级队列来实现。

3.4 贪婪最佳优先搜索

它扩展估计最接近目标的节点。它基于 $f(n) = h(n)$ 扩展节点。它使用优先级队列实现。

缺点：它可能卡在循环中。这不是最佳的。

4. 本地搜索算法

它们从一个预期的解决方案开始，然后转移到邻近的解决方案。即使它们在结束之前的任何时间被中断，它们也可以返回一个有效的解决方案。

4.1 爬山搜索

它是一种迭代算法，从问题的任意解决方案开始，并尝试通过逐步更改解决方案的单个元素来找到更好的解决方案。如果更改产生更好的解决方案，则将新增更改视为新解决方案。重复该过程直到没有进一步的改进。

函数 Hill-Climbing（problem）返回一个局部最大值的状态。

缺点：该算法既不完整也不优化。

4.2 局部集束搜索

在该算法中，它在任何给定时间保持 k 个状态。一开始，这些状态是随机生成的。这些 k 个状态的后继者是在目标函数的帮助下计算出来的。如果这些后继者中的任何一个是目标函数的最大值，则算法停止。

否则，（初始 k 状态和 k 个状态的后继者 $= 2k$）状态被放置在池中。然后按数字对池进行排序。选择最高的 k 个状态作为新的初始状态。此过程持续进行直到达到最大值。

函数 BeamSearch（problem，k）返回一个解状态。

4.3 模拟退火

退火是加热和冷却金属而改变其内部结构以改变其物理性质的过程。当金属冷却时，就形成了其新结构，而且金属保留其新获得的性能。在模拟退火过程中，温度一直可变。

我们最初将温度设置为高，然后在算法进行时让它慢慢"冷却"。当温度为高时，算法允许接受具有高频率的、更差的解决方案。

4.4 旅行商问题

在该算法中，目标是找到一个低成本旅行方案，从某一城市开始，访问途中所有城市，每个城市只过一次，最后回到起始城市。

Unit 3

录音

Text A

Best-First Search

1. Preamble

This article is about best-first search in its general form. There are a number of specific algorithms that follow the basic form of best-first search but use more sophisticated evaluation functions. A* search is a popular variant of the best-first search. This passage does not discuss A* search or other variants specifically, though the information contained herein is relevant to those search algorithms.

2. Definition

Best-first search in its most general form is a simple heuristic search algorithm. 'Heuristic' here refers to a general problem-solving rule or set of rules that do not guarantee the best solution or even any solution, but serves as a useful guide for problem-solving. Best-first search is a graph-based search algorithm, meaning that the search space can be represented as a series of nodes connected by paths.

3. How It Works

The name 'best-first' refers to the method of exploring the node with the best 'score' first. An evaluation function is used to assign a score to each candidate node. The algorithm maintains two lists, one containing a list of candidates yet to explore (OPEN), and one containing a list of visited nodes (CLOSED). Since all unvisited successor nodes of every visited node are included in the OPEN list, the algorithm is not restricted to only exploring successor nodes of the most recently visited node. In other words, the algorithm always chooses the best of all unvisited nodes that have been graphed, rather than being restricted to only a small subset, such as immediate neighbours. Other search strategies, such as depth-first and breadth-first, have this restriction. The advantage of this strategy is that if the algorithm reaches a dead-end node, it will continue to try other nodes.

4. Algorithm

Best-first search in its most basic form consists of the following algorithm.

The first step is to define the OPEN list with a single node, the starting node. The second step is to check whether or not OPEN is empty. If it is empty, then the algorithm returns failure and exits. The third step is to remove the node with the best score, n, from OPEN and place it in CLOSED. The fourth step 'expands' the node n, where expansion is the identification of successor nodes of n. The fifth step then checks each of the successor nodes to see whether or not one of them is the goal node. If any successor is the goal node, the algorithm returns success and the solution, which consists of a path traced backwards from the goal to the start node. Otherwise, the algorithm proceeds to the sixth step. For every successor node, the algorithm applies the evaluation function, f, to it, then checks to see if the node has been in either OPEN or CLOSED. If the node has not been in either, it gets added to OPEN. Finally, the seventh step establishes a looping structure by sending the algorithm back to the second step. This loop will only be broken if the algorithm returns success in step five or failure in step two.

The algorithm is represented here in pseudocode:

(1) Define a list, OPEN, consisting solely of a single node, the start node, s.

(2) IF the list is empty, return failure.

(3) Remove from the list the node n with the best score (the node where f is the minimum), and move it to a list, CLOSED.

(4) Expand node n.

(5) IF any successor to n is the goal node, return success and the solution (by tracing the path from the goal node to the start node).

(6) FOR each successor node：

- Apply the evaluation function, f, to the node.
- IF the node has not been in either list, add it to OPEN.
- Loop structure by sending the algorithm back to the second step.

Pearl adds a third step to the FOR loop that is designed to prevent reexpansion of nodes that have already been visited. This step has been omitted above because it is not common to all best-first search algorithms.

5. Evaluation Function

The particular evaluation function used to determine the score of a node is not precisely defined in the above algorithm, because the actual function used is up to the determination of the programmer, and may vary depending on the particularities of the search space. While the evaluation function can determine to a large extent the effectiveness and efficiency of the search, for the purposes of understanding the search algorithm we need not be concerned with the particularities of the function.

6. Applications

Best-first search and its more advanced variants have been used in such applications as games and web crawlers. In a web crawler, each web page is treated as a node, and all the hyperlinks on the page are treated as unvisited successor nodes. A crawler that uses best-first search generally uses an evaluation function that assigns priority to links based on how closely the contents of their parent page resemble the search query. In games, best-first search may be used as a path-finding algorithm for game characters. For example, it could be used by an enemy agent to find the location of the player in the game world. Some games divide up the terrain into 'tiles' which can either be blocked or unblocked. In such cases, the search algorithm treats each tile as a node, with the neighboring unblocked tiles being successor nodes, and the goal node being the destination tile.

✎ New Words

preamble	[prɪˈæmbl]	n. 序；绪言
variant	[ˈveərɪənt]	n. 变体,变种,变异体；变量
		adj. 不同的,相异的,不一致的；多样的；变异的

discuss	[dɪˈskʌs]	vt.讨论,谈论;论述,详述
definition	[ˌdefɪˈnɪʃn]	n.定义;解释
rule	[ruːl]	n.规则,规定;统治,支配 vi.控制,支配
guarantee	[ˌgærənˈtiː]	n.保证,担保;保证人,保证书 vt.保证,担保
score	[skɔː]	n.&v.得分;记分
candidate	[ˈkændɪdət]	n.候选者,候选人
unvisited	[ʌnˈvɪzɪtɪd]	adj.未访问的
include	[ɪnˈkluːd]	vt.包括,包含
restrict	[rɪˈstrɪkt]	vt.限制,限定;约束
immediate	[ɪˈmiːdɪət]	adj.最接近的;立即的;直接的
restriction	[rɪˈstrɪkʃn]	n.限制,限定
empty	[ˈemptɪ]	adj.空的 vt.(使)成为空的
failure	[ˈfeɪljə]	n.失败
exit	[ˈeksɪt]	n.出口,退出 vi.离开;退出
remove	[rɪˈmuːv]	vt.删除,去除
pseudocode	[ˈsjuːdəʊkəʊd]	n.伪代码
solely	[ˈsəʊllɪ]	adv.唯一地;仅仅
minimum	[ˈmɪnɪməm]	n.最小量;极小值
trace	[treɪs]	vt.跟踪,追踪;追溯
prevent	[prɪˈvent]	vt.防止,预防;阻碍;阻止 vi.阻止
reexpansion	[riːɪkˈspænʃn]	n.再扩展
omit	[əʊˈmɪt]	vt.省略;删掉
determine	[dɪˈtɜːmɪn]	vt.决定,确定
precisely	[prɪˈsaɪslɪ]	adv.精确地;恰好地
determination	[dɪˌtɜːmɪˈneɪʃn]	n.决定,确定
particularity	[pəˌtɪkjuˈlærɪtɪ]	n.特性
extent	[ɪkˈstent]	n.程度;长度
efficiency	[ɪˌfekˈtɪvnɪs]	n.有效性;有效,有力
crawler	[ˈkrɔːlə]	n.爬虫,爬行动物
hyperlink	[ˈhaɪpəlɪŋk]	n.超链接
resemble	[rɪˈzembl]	vt.与……相似,类似于
query	[ˈkwɪərɪ]	v.查询
character	[ˈkærɪktə]	n.角色,人物;性格;特点;字母

enemy	['enəmɪ]	n. 敌军
		adj. 敌人的；敌方的
terrain	[tə'reɪn]	n. 地面，地带
tile	[taɪl]	n. 片状材料，瓦片，瓷砖
		vt. 用瓦片、瓷砖等覆盖
destination	[ˌdestɪ'neɪʃn]	n. 目的，目标

✎ Phrases

refer to	涉及；指的是；适用于；参考
general problem-solving rule	普通问题解决规则
graph-based search algorithm	基于图的搜索算法
evaluation function	评价函数
dead-end node	死角节点
consist of	由……组成；包括
traced backward	追溯
web crawler	网络爬虫
web page	网页
be treated as	被当作
parent page	父页面
divide up	分割

✎ Exercises

【Ex. 1】 根据课文内容回答问题。

1. What is best-first search?
2. How many lists does the algorithm maintain? What are they?
3. What is the advantage of this strategy?
4. What is the first step?
5. What does the algorithm return if any successor is the goal node?
6. What does the seventh step do?
7. Why is the particular evaluation function used to determine the score of a node not precisely defined in the above algorithm?
8. What is each web page treated as in a web crawler? And what about the hyperlinks on the page?
9. What may best-first search be used as in games?
10. What do some games divide up the terrain into? In such cases, what does the search algorithm do?

【Ex.2】 把下列单词或词组中英互译。

1. web crawler _____   1. _____
2. web page _____   2. _____
3. graph-based search algorithm _____   3. _____
4. crawler _____   4. _____
5. definition _____   5. _____
6. n.超链接 _____   6. _____
7. n.伪代码 _____   7. _____
8. v.查询 _____   8. _____
9. n.限制,限定 _____   9. _____
10. adv.精确地;恰好地 _____   10. _____

【Ex.3】 短文翻译。

Hill Climbing

In computer science，hill climbing is a mathematical optimization technique which belongs to the family of local search. It is relatively simple to implement，making it a popular first choice. Although more advanced algorithms may give better results，in some situations hill climbing works just as well.

Hill climbing can be used to solve problems that have many solutions，some of which are better than others. It starts with a random (potentially poor) solution, and iteratively makes small changes to the solution，each time improving it a little. When the algorithm cannot see any improvement anymore，it terminates. Ideally，at that point the current solution is close to optimal，but it is not guaranteed that hill climbing will ever come close to the optimal solution.

For example，hill climbing can be applied to the traveling salesman problem. It is easy to find a solution that visits all the cities but will be very poor compared to the optimal solution. The algorithm starts with such a solution and makes small improvements to it，such as switching the order in which two cities are visited. Eventually，a much better route is obtained.

Hill climbing is used widely in artificial intelligence. It reaches a goal state from a starting node.

【Ex.4】 将下列词填入适当的位置(每个词只用一次)。

determined	heuristic	memory	advantage	recognition
heuristic	available	optimal	solved	nodes

Beam Search

1. Definition

Beam search is a restricted, or modified, version of either a breadth-first search or a best-first search. It is restricted in the sense that the amount of memory available for storing the set of alternative search __1__ is limited, and in the sense that non-promising nodes can be pruned at any step in the search. The pruning of non-promising nodes is __2__ by problem-specific heuristics. The set of most promising, or best alternative, search nodes is called the 'beam'. Essentially, beam search is a forward-pruning, __3__ search.

2. Search Components and Algorithm

A beam search takes three components as its input: a problem to be __4__, a set of heuristic rules for pruning, and a memory with a limited __5__ capacity. The problem is the problem to be solved, usually represented as a graph, and contains a set of nodes in which one or more of the nodes represents a goal. The set of __6__ rules are rules specific to the problem domain and prune unfavorable nodes from the memory in respect to the problem domain. The memory is where the 'beam' is stored, where when memory is full and a node is to be added to the beam, the most costly node will be deleted, such that the __7__ limit is not exceeded.

3. Advantages, Disadvantages, and Practical Applications

Beam search has the advantage of potentially reducing the computation, and hence the time, of a search. The memory consumption of the search is far less than its underlying search methods. This potential __8__ rests upon the accuracy and effectiveness of the heuristic rules used for pruning, and having such rules can be somewhat difficult due to the expert knowledge required of the problem domain. The main disadvantages of a beam search are that the search may not result in an __9__ goal and may not even reach a goal at all. In fact, the beam search algorithm terminates for two cases: a required goal node is reached, or a goal node is not reached and there are no nodes left to be explored. Beam search has the potential to be incomplete. Despite these disadvantages, beam search has found success in the practical areas of speech __10__, vision, planning, and machine learning.

Text B

Why AI Needs Security?

Artificial intelligence (AI) is creating new waves of innovation and business models, powered by new technology for deep learning and a massive growth in investment. As AI becomes pervasive in computing applications, high-grade security in all levels of the system is needed. Protecting AI systems, their data, and their communication is critical for users' safety and privacy, and for protecting businesses' investments.

1. Where and Why AI Security Is Needed

AI applications built around artificial neural networks operate in two basic stages—training and inference. During the training stage, a neural network 'learns' to do a job, such as recognizing faces or street signs. The resulting data set of weights, representing the strength of interaction between the artificial neurons, is used to configure the neural net as a model. In the inference stage, this model is used by the end application to infer information about the data with which it is presented.

The algorithms used in neural net training often process data, such as facial images or fingerprints, which comes from public surveillance, face recognition and fingerprint biometrics, financial or medical applications. This kind of data is usually private and often contains personally identifiable information. Attackers, whether organized crime groups or business competitors, can take advantage of this information to gain economic or other benefits.

AI systems also face the risk of being sent rogue data to disrupt a neural network's functionality. Companies that protect training algorithms and user data will be differentiated in their fields from companies that face the reputational and financial risks of not doing so. Hence, designers must ensure that data is received only from trusted sources and that it is protected during use.

The models themselves, represented by the neural-net weights that emerge during the training process, are expensive to create and form valuable intellectual property that must be protected against disclosure and misuse.

Another strong driver for enforcing personal data privacy is the General Data Protection Regulation (GDPR) that came into effect within the European Union on 25 May 2018. This legal framework sets guidelines for the collection and processing of personal information. The GDPR sets out the principles for data-management protection and the rights of the individual and large fines may be imposed on businesses that do not comply with the rules.

As data and models move between the network edge and the cloud, communications also need to be secured and authenticated. It is important to ensure that data and/or models are protected and can only be downloaded and communicated from authorized sources to authorized devices.

2. AI Security Solutions

Product security must be incorporated throughout the product lifecycle, from conceptualization to disposal. As new AI applications and use cases emerge, devices that run these applications must be able to adapt to an evolving threat landscape. To meet high-grade protection requirements, security needs to be multi-faceted and deeply embedded in everything from edge devices that use neural-network processing system-on-chips (SoCs), through the applications that run on them, to communications to the cloud and storage within it.

System designers adding security to their AI products must consider a few foundational functions for enabling security in AI products to protect all phases of operation: offline, during power up, and at runtime, including during communication with other devices or the cloud. Establishing the integrity of the system is essential to creating trust that it is behaving as intended.

2.1 Secure Bootstrap

Secure bootstrap, an example of a foundational security function, establishes that the software or firmware of the product is intact. This integrity ensures that when a product is coming out of reset, it does what its manufacturer intended—and not something that a hacker has altered it to do. Secure bootstrap systems use cryptographic signatures on the firmware to determine their authenticity.

2.2 Key Management

The best encryption algorithms can be compromised if the keys are not protected with key management, which is another foundational security function. For high-grade protection, the secret key material should reside inside a hardware root of trust.

2.3 Secure Updates

A third foundational function relates to secure updates. AI applications will continue to get more sophisticated and so data and models will need to be updated continuously. The process of distributing new models securely needs to be protected with end-to-end security. Hence it is essential that products can be updated in a trusted way to fix bugs, close vulnerabilities, and evolve product functionality. A flexible, secure update function can even be used to allow post-consumer enablement of optional features of hardware or firmware.

2.4 Protecting Data and Coefficients

After addressing foundational security issues, designers must consider how to protect the data and coefficients in their AI systems. Many neural network applications operate on audio, images, video streams, and other real-time data. There are often serious privacy concerns with these large data sets, so it is essential to protect that data when it is in working memory, or stored locally on disk or flash memory systems.

3. Conclusion

Providers of AI solutions are investing significantly in R&D, and so the neural network algorithms and the models derived from training them need to be properly protected. Concerns about the privacy of personal data, which are already being reflected in the introduction of regulations such as GDPR, also mean that it is increasingly important for companies providing AI solutions to secure them as well as possible.

✎ New Words

innovation	[ˌɪnəˈveɪʃn]	n. 改革,创新;新观念,新发明
pervasive	[pəˈveɪsɪv]	adj. 普遍的;扩大的
		adv. 无处不在地;遍布地
		n. 无处不在;遍布
critical	[ˈkrɪtɪkl]	adj. 关键的;极重要的
privacy	[ˈprɪvəsɪ]	n. 隐私,秘密
operate	[ˈɒpəreɪt]	v. 运转;操作;经营;管理
stage	[steɪdʒ]	n. 阶段
configure	[kənˈfɪgə]	v. 配置;设定
infer	[ɪnˈfɜː]	vt. 推断;猜想,推理
		vi. 作出推论
facial	[ˈfeɪʃl]	adj. 面部的;表面的
fingerprint	[ˈfɪŋgəprɪnt]	n. 指纹,指印

		vt.采指纹
surveillance	[sɜː'veɪləns]	n.监督
biometrics	[ˌbaɪəʊ'metrɪks]	n.生物测定学
identifiable	[aɪˌdentɪ'faɪəbl]	adj.可辨认的,可识别的
attacker	[ə'tækə]	n.攻击者
competitor	[kəm'petɪtə]	n.竞争者;对手
risk	[rɪsk]	n.危险,冒险
		vt.冒……的危险
functionality	[ˌfʌŋkʃə'nælɪtɪ]	n.功能性
reputational	[ˌrepjuː'teɪʃənl]	n.声誉
trusted	[trʌstɪd]	adj.可信的,无错的
weight	[weɪt]	n.权重
valuable	['væljʊəbl]	adj.贵重的,宝贵的;有价值的
disclosure	[dɪs'kləʊʒə]	n.泄露,揭露
misuse	[ˌmɪs'juːz]	vt.使用……不当;滥用
enforce	[ɪn'fɔːs]	vt.实施,执行;加强
framework	['freɪmwɜːk]	n.构架;框架
guideline	['gaɪdlaɪn]	n.指导方针;指导原则
collection	[kə'lekʃn]	n.收集,采集
authenticate	[ɔː'θentɪkeɪt]	vt.认证;鉴定
ensure	[ɪn'ʃʊə]	vt.确保
download	[ˌdaʊn'ləʊd]	v.下载
authorize	['ɔːθəraɪz]	vt.授权,批准
lifecycle	['laɪfˌsaɪkl]	n.生命周期
disposal	[dɪ'spəʊzl]	n.(事情的)处置;清理
		adj.处理废品的
threat	[θret]	n.威胁
foundational	[faʊn'deɪʃənl]	adj.基本的,基础的
offline	[ˌɒf'laɪn]	adj.未连线的;未联机的;脱机的;离线的
		adv.未连线地;未联机地;脱机地;离线地
establish	[ɪ'stæblɪʃ]	vt.建立,创建
integrity	[ɪn'tegrɪtɪ]	n.完整,完整性
essential	[ɪ'senʃl]	adj.基本的;必要的;本质的
		n.必需品;基本要素
behave	[bɪ'heɪv]	vi.表现
bootstrap	['buːtstræp]	n.引导
firmware	['fɜːmweə]	n.(计算机的)固件
intact	[ɪn'tækt]	adj.完整无缺的,未受损伤的;原封不动的;

reset	[ˌriːˈset]	vt. 重置；重排；重新安装
manufacturer	[ˌmænjʊˈfæktʃərə]	n. 制造商，制造厂
hacker	[ˈhækə]	n. (电脑)黑客
cryptographic	[ˈkrɪptəʊˈgræfɪk]	adj. 加密的，用密码写的
authenticity	[ˌɔːθenˈtɪsɪtɪ]	n. 可靠性，确实性，真实性
compromise	[ˈkɒmprəmaɪz]	n. 损害；妥协，折中方案
		vi. 折中解决；妥协
sophisticated	[səˈfɪstɪkeɪtɪd]	adj. 复杂的；精致的；富有经验的
continuously	[kənˈtɪnjʊəslɪ]	adv. 连续不断地，接连地
end-to-end	[end-tuː-end]	adj. 端到端的，端对端的
vulnerability	[ˌvʌlnərəˈbɪlɪtɪ]	n. 弱点
enablement	[ɪˈneɪblmənt]	n. 允许，启动，实现
hardware	[ˈhɑːdweə]	n. 计算机硬件
coefficient	[ˌkəʊɪˈfɪʃnt]	n. 系数
disk	[dɪsk]	n. 磁盘
reflect	[rɪˈflekt]	v. 反映，反射；考虑
introduction	[ˌɪntrəˈdʌkʃn]	n. 介绍；引言，导言

✎ Phrases

business model	企业模型，商业模式
recognizing face	识别人脸
street sign	街道标志
public surveillance	群众监督
face recognition	面貌识别
crime group	犯罪集团
financial risk	财务风险
personal data privacy	个人数据隐私
use case	用例
adapt to	使适应于
power up	加电，使启动
fix bug	修复错误
video stream	视频流
real-time data	实时数据
flash memory	闪存
derived from...	来源于……

✎Abbreviations

SoCs（System-on-Chips）   片载系统
R&D（research and development）   科学研究与开发

✎Exercises

【Ex.5】 根据课文内容填空。

1. AI applications built around _____ operate in two basic stages—training and inference. During the training stage，a neural network 'learns' to do a job，such as _____ or _____.

2. The algorithms used in neural net training often process data，such as _____ or _____，which comes from _____，face recognition and fingerprint biometrics，_____ or medical applications.

3. AI systems also face the risk of being sent _____ to disrupt a neural network's _____.

4. As data and models move between _____ and _____，communications also need to be secured and _____.

5. To meet high-grade protection requirements，security needs to be _____ and _____ in everything from edge devices that use neural-network processing system-on-chips（SoCs），through _____ that run on them，to communications to _____ and storage within it.

6. System designers adding security to their AI products must consider a few foundational functions for enabling security in AI products to protect all phases of operation：_____，_____，and _____，including during communication with other devices or the cloud.

7. Secure bootstrap，an example of a foundational security function，establishes that _____ or _____ of the product is intact.

8. For high-grade protection，the secret key material should reside _____.

9. A flexible，secure update function can even be used to allow _____ of optional features of hardware or firmware.

10. Concerns about _____ also mean that it is increasingly important for companies providing AI solutions to _____.

Reading

Heuristic Evaluation and Iterated Prisoner's Dilemma

1. Heuristic Evaluation

A heuristic evaluation is a process by which an expert evaluates a user interface[①] or similar system using a list of guidelines. This is not the same as a user evaluation or usability test[②] in which users actually try out the interface. Instead, a predetermined[③] list of features or aspects of a user interface that are commonly accepted as being beneficial[④] is used to evaluate the interface. A heuristic evaluation is typically faster and less expensive than a usability test, though it does have weaknesses and should be used early in development.

There are different ways in which a heuristic evaluation can be conducted, but it typically begins with a list of criteria or features expected of a strong user interface. This list can come from a number of sources, though the first such basic list was created by Jakob Nielsen, and establishes[⑤] 10 principle design elements that should be included in an interface. Different experts in usability and design can create their own lists, or use these 10 as a starting point for more detailed checklists[⑥]. When that expert is called upon to perform a heuristic evaluation, he or she uses the checklist to consider the strengths and weaknesses of a system.

A heuristic evaluation is usually conducted by an expert in usability features and interface design, rather than actual test users. The expert looks at the different elements of an interface and evaluates each part of it according to the checklist he or she has created. This can include the use of 'yes' or 'no' answers to evaluate if certain elements are present in the interface, as well as a numerical scale to indicate the severity of problems or issues found in the heuristic evaluation. The scale allows program developers to easily recognize the nature of a problem and quickly determine

---

① user interface：用户接口，用户界面
② usability test：可用性测试
③ predetermine [ˌpriːdɪˈtɜːmɪn] v. 预定
④ beneficial [ˌbenɪˈfɪʃl] adj. 有利的，有益的
⑤ establish [ɪˈstæblɪʃ] vt. 建立，创建
⑥ checklist [ˈtʃeklɪst] n. 清单

if the resources are available to correct it prior to software release[①].

One of the major weaknesses of a heuristic evaluation is that it applies common standards to different types of systems. A feature that may be required in one type of software may be unnecessary[②] in another; while some features that might be considered poor design for some programs can actually be beneficial in others. Many companies still utilize experts to perform a heuristic evaluation, however, since the process is faster and cheaper than long-term usability testing using large groups of users. Heuristic evaluations are still beneficial, but they should be used early on in the design and development process so that changes suggested by the evaluation can be considered prior to usability testing that often demonstrates[③] the reality of interface usability.

2. Iterated Prisoner's Dilemma

2.1 Definition of Iterated Prisoner's Dilemma[④]

The iterated prisoner's dilemma is an extension of the general form except the game is repeatedly played by the same participants. An iterated prisoner's dilemma differs from the original concept of a prisoner's dilemma because participants can learn about the behavioral tendencies of their counterparty[⑤].

The iterated prisoner's dilemma at times has been called the 'Peace-War game'.

2.2 Breaking down Iterated Prisoner's Dilemma

Since the game is repeated, one individual can formulate a strategy that does not follow the regular logical convention of an isolated round.

'Tit for tat[⑥]' is a common iterated prisoner's dilemma strategy.

The iterated prisoner's dilemma game is fundamental to many theories of human cooperation[⑦] and trust. Based on the assumption[⑧] that the game can model transactions between two people requiring trust, cooperative behavior in populations[⑨]

---

① release [rɪ'liːs] vt. 发布,发行

② unnecessary [ʌn'nesəsərɪ] adj. 不必要的,多余的;无用的

③ demonstrate ['demənstreɪt] vt. 证明,证实;论证;显示

④ iterated prisoner's dilemma: 迭代囚徒困境

⑤ counterparty ['kaʊntə'pɑːtɪ] n. 对手方

⑥ tit for tat: 针锋相对;以牙还牙,一报还一报

⑦ cooperation [kəʊˌɒpə'reɪʃn] n. 合作,协作

⑧ assumption [ə'sʌmpʃn] n. 假定,假设

⑨ population [ˌpɒpjuˈleɪʃn] n. 人口

may be modeled by a multi-player, iterated, version of the game.

The theory behind the game has captivated① many scholars② over the years. More recently, organizational design researchers have used the game to model corporate strategies. The prisoner's dilemma is also now commonplace③ for game theories are becoming popular with investment strategist. Globalization④ and integrated trade have further driven demand for financial and operational models that can describe geopolitical⑤ issues.

2.3   Example of the Iterated Prisoner's Dilemma Game

For example, you and a colleague are in jail and suspected of committing a crime. You are isolated⑥ from each other and do not know how the other will respond to questioning. The police invite both of you to implicate the other in the crime (defect). What happens depends on what both of you do, but neither of you know how the other will respond. If your colleague betrays you (yields to the temptation to defect) while you remain silent⑦, then you receive the longest jail term⑧ while your colleague gets off free (and visa versa). If you both choose to cooperate with each other (not the police) by remaining silent, there is insufficient⑨ evidence⑩ to convict⑪ both of you, so you are both given a light sentence for a lesser crime. If both of you decide to defect, you have condemned each other to slightly reduced but still heavy sentences.

The game is played iteratively for a number of rounds until it is ended (as if you are repeatedly⑫ interrogated⑬ for separate crimes). The scores from each round are accumulated, so the object is to optimize the point score before reaching game over. Game over is determined randomly anywhere between 1 and 100 rounds. At the end of the game, the scores are translated into percentages of the best possible scores.

---

① captivate ['kæptɪveɪt] *vt*. 迷住,迷惑

② scholar ['skɒlə] *n*. 学者

③ commonplace ['kɒmənpleɪs] *adj*. 平凡的,普通的

④ globalization [ˌgləʊbəlaɪ'zeɪʃn] *n*. 全球化,全球性

⑤ geopolitical [ˌdʒiːəʊpə'lɪtɪkl] *adj*. 地理政治学的

⑥ isolate ['aɪsəleɪt] *vt*. 使隔离,使孤立

⑦ remain silent:保持沉默,沉默无语

⑧ jail term:刑期

⑨ insufficient [ˌɪnsə'fɪʃnt] *adj*. 不足的,不够的

⑩ evidence ['evɪdəns] *n*. 证词,证据

⑪ convict [kən'vɪkt] *vt*. 宣判有罪 *n*. 罪犯

⑫ repeatedly [rɪ'piːtɪdlɪ] *adv*. 反复地,重复地;屡次地

⑬ interrogate [ɪn'terəgeɪt] *vt*. 审问

参考译文

最佳优先搜索

1. 序言

本文介绍了最佳优先搜索的一般形式。有许多特定的算法遵循最佳优先搜索的基本形式，却使用更复杂的评估函数。A*搜索是最佳优先搜索的一个流行变体。虽然此处包含的信息与这些搜索算法相关，但本文没有专门讨论 A*搜索或其他变体。

2. 定义

最佳优先搜索的最常见形式是简单的启发式搜索算法。这里的"启发式"是指一般的问题解决规则或一组规则，它们不能保证得到最佳解决方案甚至不能保证得到任何解决方案，但可以作为解决问题的有用指南。最佳优先搜索是基于图的搜索算法，这意味着搜索空间可以表示为由路径连接的一系列节点。

3. 它如何工作

名称"最佳优先"指首先探索具有最佳"得分"的节点的方法。用评估函数为每个候选节点分配分数。该算法含有两个列表，一个列表包含尚待探索的候选节点（OPEN），另一个列表包含已访问节点（CLOSED）。由于每个访问节点的所有未访问的后继节点都包括在 OPEN 列表中，因此该算法不限于仅探索最近访问的节点的后继节点。换句话说，该算法总是选择图中所有未访问节点中的最佳节点，而不是仅限于小的子集（例如直接邻居）。其他搜索策略（例如深度优先和广度优先）具有这样的限制。该策略的优点是，如果算法到达死角节点，它将继续尝试其他节点。

4. 算法

最佳优先搜索最基本形式包括以下算法。

第一步是使用单个节点（起始节点）定义 OPEN 列表。第二步是检查 OPEN 是否为空。如果它为空，则算法返回失败并退出。第三步是从 OPEN 中删除具有最高分数 n 的节点，并将其置于 CLOSED 中。第四步"扩展"节点 n，其中扩展是 n 的后继节点的标识。然后，第五步检查每个后继节点，以查看它们中的一个是否是目标节点。如果任何后继者是目标节点，则算法返回成功和解决方案，该解决方案包括从目标向后追溯到起始节点的路径。否则,算法前进到第六步。对于每个后继节点,算法将评估函数 f 应用于它,然后检查节点是否已处于 OPEN 或 CLOSED 状态。如果该节点尚未进入任一列表,则会将其添加到 OPEN。最后,第七步通过将算法回送到第二步来建立循环结构。如果算法在第五步中返回成功或在第二步中失败,则该循环中断。

这里的算法用伪代码表示为:

(1) 定义一个列表,OPEN,仅由单个节点(即起始节点)组成。

(2) 如果列表为空,则返回失败。

(3) 从列表中删除具有最佳分数的节点 n(f 为最小的节点),并将其移至列表 CLOSED。

(4) 展开节点 n。

(5) 如果 n 的任何后继者是目标节点,则返回成功和解决方案(通过跟踪从目标节点到开始节点的路径)。

(6) 对于每个后继节点:

- 将评估函数 f 应用于节点。
- 如果节点未在任何列表中,则将其添加到 OPEN。
- 通过将算法回送到第二步来建立循环结构。

Pearl 为 FOR 循环添加了第三步,旨在防止重新扩展已经访问过的节点。上面省略了该步骤,因为对于所有最佳优先搜索算法这一步并不常见。

5. 评估函数

用于确定节点得分的特定评估函数在上述算法中没有精确定义,因为所使用的实际函数由程序员确定,并且可以根据搜索空间的特性而变化。虽然评估函数可以在很大程度上确定搜索的有效性和效率,但是为了理解搜索算法,我们不需要关心函数的特性。

6. 应用

最佳搜索及其更高级的变体已用于游戏和网络爬虫等应用程序。在网络爬虫程序中,每个网页都被视为一个节点,页面上的所有超链接都被视为未访问的后继节点。使用最佳优先搜索的网络爬虫程序通常使用评估函数,该函数根据其父页面的内容与搜索查询的相似程度为链接分配优先级。在游戏中,最佳优先搜索可以为游戏角色提供路径搜索算法。例如,对手可以使用它来查找游戏世界中玩家的位置。有些游戏把地形分成可进入或不可进入的区块。在这种情况下,搜索算法将每个区块视为节点,其中相邻的可进入区块是后继节点,并且目标节点是目标区块。

Unit 4

录音

Text A

Computer Software

Computer software, consisting of programs, enables a computer to perform specific tasks. It is opposed to its physical components (hardware) which can only do the tasks they are mechanically designed for. The term includes application software such as word processors, which perform productive tasks for users, system software such as operating systems, which interface with hardware to run the necessary services for user-interfaces and applications, and middleware, which controls and coordinates distributed systems.

1. Terminology

The term 'software' is an instruction-procedural programming source for scheduling instruction streams according to the von Neumann machine paradigm. It should not be confused with Configware and Flowware, which are programming sources for configuring the resources (structural 'programming' by Configware) and for scheduling the data streams (data-procedural programming by Flowware) of the Anti machine paradigm of Reconfigurable Computing systems.

2. Relationship to Computer Hardware

Computer software is so called in contrast to computer hardware, which encompasses the physical interconnections and devices required to store and execute (or run) the software. In computers, software is loaded into RAM and executed in the central processing unit. At the lowest level, software consists of a machine language specific to an individual processor. A machine language consists of groups of binary values signifying processor instructions (object codes), which change the state of the computer from its preceding state. Software is an ordered sequence of instructions for changing the state of the computer hardware in a particular sequence. It is usually written in high-level programming languages that are easier and more efficient for humans to use (closer to natural language) than machine language. High-level languages are compiled or interpreted into machine language object code. Software may also be written in an assembly language, essentially, a mnemonic representation of a machine language using a natural language alphabet. Assembly language must be assembled into object code via an assembler.

In computer science and software engineering, computer software is all computer programs. The concept of reading different sequences of instructions into the memory of a device to control computations was invented by Charles Babbage as part of his difference engine.

3. Types

Practical computer systems divide software systems into three major classes: system software, programming software and application software, although the distinction is arbitrary, and often blurred.

3.1 System Software

System software helps run the computer hardware and computer system. It includes operating systems, device drivers, diagnostic tools, servers, windowing systems, utilities and more. The purpose of systems software is to insulate the applications programmer as much as possible from the details of the particular computer complex being used, especially memory and other hardware features, and such accessory devices as communications, printers, readers, displays, keyboards, etc.

3.2 Programming Software

Programming software usually provides tools to assist a programmer in writing computer programs and software using different programming languages in a more convenient way. The tools include text editors, compilers, interpreters, linkers, debuggers, and so on. An integrated development environment (IDE) merges those tools into a software bundle, and a programmer may not need to type multiple commands for compiling, interpreting, debugging, tracing, and etc., because the IDE usually has an advanced graphical user interface, or GUI.

3.3　Application Software

Application software allows end users to accomplish one or more specific (non-computer related) tasks. Typical applications include industrial automation, business software, educational software, medical software, databases, and computer games. Businesses are probably the biggest users of application software, but almost every field of human activity now uses some form of application software. It is used to automate all sorts of functions.

4. Three Layers

Users often see things differently than programmers. People who use modern general purpose computers (as opposed to embedded systems, analog computers, supercomputers, etc.) usually see three layers of software performing a variety of tasks: platform, application, and user software.

4.1　Platform Software

Platform includes the firmware, device drivers, an operating system, and typically a graphical user interface which, in total, allows a user to interact with the computer and its peripherals (associated equipment). Platform software often comes bundled with the computer, and users may not realize that it exists or that they have a choice to use different platform software.

4.2　Application Software

Application software or applications are what most people think of when they think of software. Typical examples include office suites and video games. Application software is often purchased separately from computer hardware. Sometimes applications are bundled with the computer, but that does not change the fact that they run as independent applications. Applications are almost always independent programs from the operating system, though they are often tailored for specific platforms. Most users think of compilers, databases, and other 'system software' as applications.

4.3   User Software

User software tailors systems to meet the users specific needs. User software includes spreadsheet templates, word processor macros, scientific simulations, and scripts for graphics and animations. Even email filters are a kind of user software. Users create this software themselves and often overlook how important it is. Depending on how competently the user-written software has been integrated into purchased application packages, many users may not be aware of the distinction between the purchased packages, and what has been added by fellow co-workers.

5. Operation

Computer software has to be 'loaded' into the computer's storage (such as a hard drive, memory, or RAM). Once the software is loaded, the computer is able to execute the software. Computers operate by executing the computer program. This involves passing instructions from the application software, through the system software, to the hardware which ultimately receives the instruction as machine code. Each instruction causes the computer to carry out an operation—moving data, carrying out a computation, or altering the control flow of instructions.

Data movement is typically from one place in memory to another. Sometimes it involves moving data between memory and registers which enable high-speed data access in the CPU. Moving data, especially large amounts of it, can be costly. So, this is sometimes avoided by using 'pointers' to data instead. Computations include simple operations such as incrementing the value of a variable data element. More complex computations may involve many operations and data elements together.

Instructions may be performed sequentially, conditionally, or iteratively. Sequential instructions are those operations that are performed one after another. Conditional instructions are performed such that different sets of instructions execute depending on the value(s) of some data. In some languages this is known as an 'if' statement. Iterative instructions are performed repetitively and may depend on some data value. This is sometimes called a 'loop'. Often, one instruction may 'call' another set of instructions that are defined in some other program or module. When more than one computer processor is used, instructions may be executed simultaneously.

A simple example of the way software operates is what happens when a user selects an entry such as 'Copy' from a menu. In this case, a conditional instruction is executed to copy text from data in a 'document' area residing in memory, perhaps to an intermediate storage area known as a 'clipboard' data area. If a different menu entry such as 'Paste' is chosen, the software may execute the instructions to copy the text from the clipboard data area to a specific location in the same or another document in memory.

Depending on the application, even the example above could become complicated. The field of software engineering endeavors to manage the complexity of how software operates. This is especially true for software that operates in the context of a large or powerful computer system.

Kinds of software by operation: computer program as executable, source code or script, configuration.

6. Reliability and Quality

Software reliability considers the errors, faults, and failures related to the creation and operation of software.

Software quality is very important, especially for commercial and system software like Microsoft Office, Microsoft Windows and Linux. If software is faulty (buggy), it can delete a person's work, crash the computer and do other unexpected things. Faults and errors are called 'bugs' which are often discovered during alpha and beta testing. Software is often also a victim to what is known as software aging, the progressive performance degradation resulting from a combination of unseen bugs.

Many bugs are discovered and eliminated (debugged) through software testing. However, software testing rarely—if ever—eliminates every bug; some programmers say that 'every program has at least one more bug'. In the waterfall method of software development, separate testing teams are typically employed, but in newer approaches, collectively termed agile software development, developers often do all their own testing, and demonstrate the software to users/clients regularly to obtain feedback. Software can be tested through unit testing, regression testing and other methods, which are done manually, or most commonly, automatically, since the amount of code to be tested can be quite large. For instance, NASA has extremely rigorous software testing procedures for many operating systems and communication functions. Many NASA-based operations interact and identify each other through command programs. This enables many people who work at NASA to check and evaluate functional systems overall. Programs containing command software enable hardware engineering and system operations to function much easier together.

✎ New Words

mechanically	[məˈkænɪkəlɪ]	*adv*.机械地
middleware	[ˈmɪdlweə]	*n*.中间设备，中间件

procedural	[prəˈsiːdʒərəl]	*adj*.程序上的
paradigm	[ˈpærədaɪm]	*n*.范例
structural	[ˈstrʌktʃərəl]	*adj*.结构的,结构化
interconnection	[ˌɪntəkəˈnekʃn]	*n*.互相连接
compile	[kəmˈpaɪl]	*vt*.编译
assembler	[əˈsemblə]	*n*.汇编程序
arbitrary	[ˈɑːbɪtrərɪ]	*adj*.武断的,专断的
blur	[blɜː]	*v*.模糊
insulate	[ˈɪnsjʊleɪt]	*vt*.隔离,使绝缘
reader	[ˈriːdə]	*n*.读卡机
convenient	[kənˈviːnɪənt]	*adj*.便利的,方便的
interpreter	[ɪnˈtɜːprɪtə]	*n*.解释程序
linker	[ˈlɪŋkə]	*n*.(目标代码)连接器
debugger	[ˌdiːˈbʌgə]	*n*.调试器
merge	[mɜːdʒ]	*v*.合并,并入,融合
bundle	[ˈbʌndl]	*n*.捆,束,包
		v.捆扎
tailored	[ˈteɪləd]	*adj*.定做的,特制的,专门的
template	[ˈtempleɪt]	*n*.模板(= templet)
macro	[ˈmækrəʊ]	*n*.宏
script	[skrɪpt]	*n*.脚本
animation	[ˌænɪˈmeɪʃn]	*n*.动画
filter	[ˈfɪltə]	*n*.过滤器,滤波器
competently	[ˈkɒmpɪtəntlɪ]	*adv*.胜任地,适合地
co-worker	[ˈkəʊwɜːkə]	*n*.合作者,同事,帮手
alter	[ˈɔːltə]	*v*.改变
costly	[ˈkɒstlɪ]	*adj*.昂贵的,困难的;造成损失的
pointer	[ˈpɔɪntə]	*n*.指针
conditionally	[kənˈdɪʃənəlɪ]	*adv*.有条件的
iteratively	[ˈɪtəˌreɪtɪvlɪ]	*adv*.反复地;迭代地
call	[kɔːl]	*n*.&v.调用
clipboard	[ˈklɪpbɔːd]	*n*.剪贴板
endeavor	[ɪnˈdevə]	*n*.&vi.尽力,努力
reliability	[rɪˌlaɪəˈbɪlɪtɪ]	*n*.可靠性

✎Phrases

| be opposed to | 与……相对,和……相反 |

system software	系统软件
distributed system	分布式的计算机系统
be confused with	混淆
in contrast to	和……形成对比,和……形成对照
machine language	机器语言
object code	结果代码
ordered sequence	有序序列
high-level programming language	高级编程语言
natural language	自然语言
assembly language	汇编语言
software engineering	软件工程
difference engine	差分机
divide…into…	把……分成……
device driver	设备驱动程序
diagnostic tool	诊断工具
as much as possible	尽可能
computer complex	计算装置
text editor	文本编辑器
integrated development environment (IDE)	集成开发环境
computer game	计算机游戏程序
all sorts of	各种各样的
embedded system	嵌入式系统
analog computer	模拟计算机
a variety of	多种的
platform software	平台软件
in total	整个地(＝as a whole)
come with	伴随……发生,与……一起供给
video game	计算机视频游戏,电视游戏
separate from	分离,分开
be integrated into…	统一到……中,整合到……中
be aware of	知道
be able to	能,会
carry out	完成,实现,执行
one after another	接连地
be known as	被认为是
conditional instruction	条件指令
be incapable of	不能
source code	源编码,原代码,源程序

software reliability   软件可靠性

✎ Exercises

【Ex. 1】 根据课文内容回答问题。

1. What does computer software consist of? What does it do?

2. What does the term computer software include?

3. What does a machine language consist of?

4. What are the three major classes practical computer systems divide software systems into?

5. What does system software do? What does it include?

6. What does programming software usually do?

7. What does application software usually do? What do typical applications include?

8. What are the three layers of software?

9. What are the sequential instructions, conditional instructions and iterative instructions respectively?

10. What can happen if software is faulty (buggy)?

【Ex. 2】 把下列单词或词组中英互译。

1. assembler _____   1. _____
2. compile _____   2. _____
3. filter _____   3. _____
4. interconnection _____   4. _____
5. interpreter _____   5. _____
6. adv. 反复地；迭代地 _____   6. _____
7. n. 范例 _____   7. _____
8. 汇编语言 _____   8. _____
9. 分布式的计算机系统 _____   9. _____
10. 高级编程语言 _____   10. _____

【Ex. 3】 短文翻译。

Computer programs (also software programs, or just programs) are instructions for a computer. A computer requires programs to function, typically executing the program's instructions in a central processor. The program has an executable form that the computer can use directly to execute the instructions.

Computer source code is often written by professional computer programmers. Source code may be converted into an executable file (sometimes called an executable program or a binary) by a compiler. Alternatively, computer programs may be executed by a central processing unit with the aid of an interpreter, or may be embedded directly into hardware.

Computer programs may be categorized along functional lines: system software and application software. And many computer programs may run simultaneously on a single computer, a process known as multitasking.

【Ex.4】 将下列词填入适当的位置(每个词只用一次)。

application	computer	create	enable	transferring
programs	software	user	basics	processing

System software is closely related to, but distinct from operating system software. It is any computer software that provides the infrastructure over which __1__ can operate, i.e. it manages and controls computer hardware so that __2__ software can perform. Operating systems, such as GNU, Microsoft Windows, Mac OS X or Linux, are prominent examples of system __3__ .

System software is software that basically allows the parts of a __4__ to work together. Without the system software the computer cannot operate as a single unit. In contrast to system software, software that allows you to do things like __5__ text documents, play games, listen to music, or surf the web.

In general, application programs are software that __6__ the end-user to perform specific, productive tasks, such as word __7__ or image manipulation. System software performs tasks like __8__ data from memory to disk, or rendering text onto a display device.

System software is not generally what a user would buy a computer for, instead, it is usually the __9__ of a computer which come built-in. Application software is the programs on the computer when the __10__ buys it. These may include word processors and web browsers .

Text B

Python Programming Language

Python is a widely used high-level, general-purpose, interpreted, dynamic programming language. Its design philosophy emphasizes code readability, and its syntax allows programmers to express concepts in fewer lines of code than possible in languages such as C++ or Java. The language provides constructs intended to enable clear programs on both a small and large scale.

Python supports multiple programming paradigms, including object-oriented, imperative and functional programming or procedural styles. It features a dynamic type system and automatic memory management and has a large and comprehensive standard library.

Python interpreters are available for many operating systems, allowing Python code to run on a wide variety of systems. Using third-party tools, such as Py2exe or Pyinstaller, Python code can be packaged into stand-alone executable programs for some of the most popular operating systems, so Python-based software can be distributed to, and used on, those environments with no need to install a Python interpreter.

1. Features and philosophy

Python is a multi-paradigm programming language: object-oriented programming and structured programming are fully supported, and many language features support functional programming and aspect-oriented programming. Many other paradigms are supported via extensions, including design by contract and logic programming.

Rather than requiring all desired functionality to be built into the language's core, Python was designed to be highly extensible. Python can also be embedded in existing applications that need a programmable interface.

2. Syntax and semantics

Python is intended to be a highly readable language. It is designed to have an uncluttered visual layout, often using English keywords where other languages use punctuation. Furthermore, Python has fewer syntactic exceptions and special cases than C or Pascal.

2.1 Indentation

Python uses whitespace indentation, rather than curly braces or keywords, to delimit blocks; this feature is also termed the off-side rule. An increase in indentation comes after certain statements; a decrease in indentation signifies the end of the current block.

2.2 Statements and control flow

Python's statements include (among others):

(1) The assignment statement (token ' = ', the equals sign). This operates differently than in traditional imperative programming languages, and this fundamental mechanism (including the nature of Python's version of variables) illuminates many other features of the language. Assignment in C, e. g., $x = 2$,

translates to 'typed variable name x receives a copy of numeric value 2'. The (right-hand) value is copied into an allocated storage location for which the (left-hand) variable name is the symbolic address. The memory allocated to the variable is large enough (potentially quite large) for the declared type. In the simplest case of Python assignment, using the same example, $x = 2$, translates to '(generic) name x receives a reference to a separate, dynamically allocated object of numeric (int) type of value 2.' This is termed binding the name to the object. Since the name's storage location doesn't contain the indicated value, it is improper to call it a variable. Names may be subsequently rebound at any time to objects of greatly varying types, including strings, procedures, complex objects with data and methods, etc. Successive assignments of a common value to multiple names, e.g., $x = 2$; $y = 2$; $z = 2$ result in allocating storage to (at most) three names and one numeric object, to which all three names are bound. Since a name is a generic reference holder it is unreasonable to associate a fixed data type with it. However at a given time a name will be bound to some object, which will have a type; thus there is dynamic typing.

(2) The if statement, which conditionally executes a block of code, along with else and elif (a contraction of else-if).

(3) The for statement, which iterates over an iterable object, capturing each element to a local variable for use by the attached block.

(4) The while statement, which executes a block of code as long as its condition is true.

(5) The try statement, which allows exceptions raised in its attached code block to be caught and handled by except clauses; it also ensures that clean-up code in a finally block will always be run regardless of how the block exits.

(6) The class statement, which executes a block of code and attaches its local namespace to a class, for use in object-oriented programming.

(7) The def statement, which defines a function or method.

(8) The with statement (from Python 2.5), which encloses a code block within a context manager (for example, acquiring a lock before the block of code is run and releasing the lock afterwards, or opening a file and then closing it), allowing Resource Acquisition Is Initialization (RAII)-like behavior.

(9) The pass statement, which serves as a NOP. It is syntactically needed to create an empty code block.

(10) The assert statement, used during debugging to check for conditions that ought to apply.

(11) The yield statement, which returns a value from a generator function. From Python 2.5, yield is also an operator. This form is used to implement coroutines.

(12) The import statement, which is used to import modules whose functions or

variables can be used in the current program.

(13) The print statement was changed to the print() function in Python 3.

2.3 Expressions

Some Python expressions are similar to languages such as C and Java, while some are not:

(1) Addition, subtraction, and multiplication are the same, but the behavior of division differs. Python also added the ** operator for exponentiation.

(2) As of Python 3.5, it supports matrix multiplication directly with the @operator, versus C and Java, which implement these as library functions. Earlier versions of Python also used methods instead of an infix operator.

(3) In Python, = = compares by value, versus Java, which compares numerics by value and objects by reference. (Value comparisons in Java on objects can be performed with the equals() method.) Python's is operator may be used to compare object identities (comparison by reference). In Python, comparisons may be chained, for example $a <= b <= c$.

(4) Python uses the words and, or, not for its boolean operators rather than the symbolic &&, ||, ! used in Java and C.

(5) Python has a type of expression termed a list comprehension. Python 2.4 extended list comprehensions into a more general expression termed a generator expression.

(6) Anonymous functions are implemented using lambda expressions; however, these are limited in that the body can only be one expression.

(7) Conditional expressions in Python are written as x if c else y (different in order of operands from the $c ? x : y$ operator common to many other languages).

(8) Python makes a distinction between lists and tuples. Lists are written as [1, 2, 3], are mutable, and cannot be used as the keys of dictionaries (dictionary keys must be immutable in Python). Tuples are written as (1, 2, 3), are immutable and thus can be used as the keys of dictionaries, provided all elements of the tuple are immutable. The parentheses around the tuple are optional in some contexts. Tuples can appear on the left side of an equal sign; hence a statement like $x, y = y, x$ can be used to swap two variables.

(9) Python has a 'string format' operator %. This functions analogous to printf format strings in C.

(10) Python has various kinds of string literals.

- Strings delimited by single or double quote marks. Unlike in UNIX shells, Perl and Perl-influenced languages, single quote marks and double quote marks function identically. Both kinds of string use the backslash (\) as an escape character and there is no implicit string interpolation such as ' $ spam'.

- Triple-quoted strings, which begin and end with a series of three single or double quote marks. They may span multiple lines and function like here documents in shells, Perl and Ruby.
- Raw string varieties, denoted by prefixing the string literal with an r. No escape sequences are interpreted; hence raw strings are useful where literal backslashes are common, such as regular expressions and Windows-style paths. Compare '@-quoting' in C♯.

(11) Python has array index and array slicing expressions on lists, denoted as a [key], a[start:stop] or a[start:stop:step]. Indexes are zero-based, and negative indexes are relative to the end. Slices take elements from the start index up to, but not including, the stop index. The third slice parameter, called step or stride, allows elements to be skipped and reversed. Slice indexes may be omitted, for example a[:] returns a copy of the entire list. Each element of a slice is a shallow copy.

In Python, a distinction between expressions and statements is rigidly enforced, in contrast to languages such as Common Lisp, Scheme, or Ruby. This leads to duplicating some functionality.

Statements cannot be a part of an expression, so list and other comprehensions or lambda expressions, all being expressions, cannot contain statements. A particular case of this is that an assignment statement such as $a = 1$ cannot form part of the conditional expression of a conditional statement. This has the advantage of avoiding a classic C error of mistaking an assignment operator = for an equality operator = = in conditions: if ($c = 1$) {...} is valid C code but if $c = 1$: ... causes a syntax error in Python.

2.4 Mathematics

Python has the usual C arithmetic operators (+ , − , ∗ , / , %). It also has ∗∗ for exponentiation, e.g. $5 ∗∗ 3 = = 125$ and $9 ∗∗ 0.5 = = 3.0$, and a new matrix multiply @ operator is included in version 3.5.

Python provides a round function for rounding a float to the nearest integer. For tie-breaking, versions before 3 use round-away-from-zero: round(0.5) is 1.0, round (− 0.5) is − 1.0. Python 3 uses round to even: round(1.5) is 2, round(2.5) is 2.

Python allows boolean expressions with multiple equality relations in a manner that is consistent with general use in mathematics. For example, the expression $a < b < c$ tests whether a is less than b and b is less than c. C-derived languages interpret this expression differently: in C, the expression would first evaluate $a < b$, resulting in 0 or 1, and that result would then be compared with c.

Due to Python's extensive mathematics library, it is frequently used as a scientific scripting language to aid in problems such as numerical data processing and manipulation.

3. Libraries

Python has a large standard library, commonly cited as one of Python's greatest strengths, providing tools suited to many tasks. For Internet-facing applications, many standard formats and protocols (such as MIME and HTTP) are supported. Modules for creating graphical user interfaces, connecting to relational databases, pseudorandom number generators, arithmetic with arbitrary precision decimals, manipulating regular expressions, and doing unit testing are also included.

The standard library is not needed to run Python or embed it in an application. For example, Blender 2.49 omits most of the standard library.

As of January 2016, the Python Package Index, the official repository of third-party software for Python, contains more than 72,000 packages offering a wide range of functionality, including:

(1) graphical user interfaces, web frameworks, multimedia, databases, networking and communications

(2) test frameworks, automation and web scraping, documentation tools, system administration

(3) scientific computing, text processing, image processing

4. Development environments

Most Python implementations can function as a command line interpreter, for which the user enters statements sequentially and receives the results immediately (read-eval-print loop (REPL)). In short, Python acts as a command-line interface or shell.

Other shells add abilities beyond those in the basic interpreter, including IDLE and IPython. While generally following the visual style of the Python shell, they implement features like auto-completion, session state retention, and syntax highlighting.

In addition to standard desktop integrated development environments (Python IDEs), there are also web browser-based IDEs, Sage (intended for developing science and math-related Python programs), and a browser-based IDE and hosting environment, PythonAnywhere.

🖎 New Words

high-level	[ˈhaɪ-ˈlevəl]	adj.高级的
general-purpose	[ˈdʒenrəl-ˈpɜːpəs]	adj.多种用途的
readability	[ˌriːdəˈbɪlɪtɪ]	n.易读,可读性
programmer	[ˈprəʊɡræmə]	n.程序员
imperative	[ɪmˈperətɪv]	n.命令
		adj.命令的
unclutter	[ʌnˈklʌtə]	vt.使整洁,整理
exception	[ɪkˈsepʃn]	n.异常,例外
indentation	[ˌɪndenˈteɪʃn]	n.缩排
keyword	[ˈkiːwɜːd]	n.关键字
statement	[ˈsteɪtmənt]	n.语句
declared	[dɪˈkleəd]	adj.声明的
improper	[ɪmˈprɒpə]	adj.不适当的,不合适的,不正确的
string	[strɪŋ]	n.串
successive	[səkˈsesɪv]	adj.连续的
contraction	[kənˈtrækʃn]	n.缩写式,紧缩
iterate	[ˈɪtəreɪt]	vt.重复
clause	[klɔːz]	n.子句
namespace	[ˈneɪmspeɪs]	n.名空间
class	[klɑːs]	n.类
enclose	[ɪnˈkləʊz]	vt.封装
manager	[ˈmænɪdʒə]	n.管理器
debugging	[ˈdiːˈbʌɡɪŋ]	n.调试
generator	[ˈdʒenəreɪtə]	n.生成器
coroutine	[kəruːˈtiːn]	n.协同程序
expression	[ɪkˈspreʃn]	n.表达式
exponentiation	[ˌekspəʊˌnenʃɪˈeɪʃən]	n.求幂
infix	[ˈɪnfɪks]	n.中缀
		vt.让……插进
mutable	[ˈmjuːtəbl]	adj.可变的,易变的
immutable	[ɪˈmjuːtəbl]	adj.不可变的,不能变的
tuple	[tʌpl]	n.元组
parentheses	[pəˈrenθəsiːz]	n.圆括号
optional	[ˈɒpʃənl]	adj.可选择的
analogous	[əˈnæləɡəs]	adj.类似的,相似的

literal	['lɪtərəl]	*adj.*文字的,照字面上的
backslash	['bækslæʃ]	*n.*反斜线符号
prefix	['priːfɪks]	*n.*前缀
reversed	[rɪ'vɜːst]	*v.*翻转,颠倒
rounding	['raʊndɪŋ]	*n.*舍入,取整
module	['mɒdjuːl]	*n.*模块
pseudorandom	[psjuːdəʊ'rændəm]	*adj.*伪随机的
multimedia	[ˌmʌltɪ'miːdɪə]	*n.*多媒体
communication	[kəˌmjuːnɪ'keɪʃn]	*n.*通信

✎ Phrases

dynamic programming language	动态编程语言
design philosophy	设计原理
lines of code	代码行
memory management	内存管理
standard library	标准库
third-party tool	第三方工具
structured programming	结构化编程
aspect-oriented programming	面向切面编程
design by contract	契约设计
logic programming	逻辑编程
special case	特殊情况
curly braces	大括号,花括号,大括弧
off-side rule	越位规则
control flow	控制流
assignment statement	赋值语句
symbolic address	符号地址
storage location	存储位置,存储单元
matrix multiplication	矩阵乘法
library function	库函数
object identity	对象标识
boolean operator	布尔运算符,逻辑运算符
list comprehension	列表解析,列表推导
generator expression	生成器表达式
anonymous function	匿名函数
lambda expression	λ表达式
conditional expression	条件表达式

make a distinction between...	对……加以区别
string literal	字符串字面量
single quote mark	单引号
double quote mark	双引号
escape character	转义字符
string interpolation	字符串插值
triple-quoted string	三重引号字符串
regular expression	正则表达式
array index	数组下标
array slicing	数组切片
shallow copy	浅拷贝
particular case	特别情况,特例
have the advantage of	胜过
boolean expression	逻辑表达式,布尔表达式
in a manner	在某种意义上
pseudorandom number generator	伪随机数产生器
web scraping	网络爬虫,网络数据抓取
development environment	开发环境
command line interpreter	命令行解释程序
session state retention	会话状态保留
hosting environment	托管环境

Abbreviations

RAII（Resource Acquisition Is Initialization）	资源获得即初始化
NOP（No Operation）	无操作
MIME（Multipurpose Internet Mail Extensions）	多用途互联网邮件扩展
HTTP（HyperText Transfer Protocol）	超文本传输协议
REPL（read-eval-print loop）	读取—求值—打印循环

Exercises

【Ex.5】 根据课文内容回答问题。

1. What is Python?

2. What does the if statement conditionally do?

3. What does the try statement do?

4. What does Python 3.5 support matrix multiplication with?

5. What distinction does Python make between lists and tuples?

6. What do triple-quoted strings begin and end with?

7. What is the third slice parameter called? What does it do?

8. What is one of Python's greatest strengths?

9. What can most Python implementations function as?

Reading

5 Most Popular AI Programming Languages

Just like in the development of most software applications, a developer has a variety of languages to use in writing AI. However, there is no perfect programming language to point as the best programming language used in artificial intelligence. The development process depends on the desired functionality[①] of the AI application being developed. AI has so far achieved biometric intelligence, autopilots[②] for self-driving cars and other applications that required different artificial intelligence coding language for their development projects.

Java, Python, Lisp, Prolog, and C++ are major AI programming language used for artificial intelligence capable of satisfying different needs in the development and designing of different software. It is up to a developer to choose which of the AI languages will gratify[③] the desired functionality and features of the application requirements.

1. Python

Python is a favorite programming language in AI development because of its syntax simplicity and versatility[④]. Python is very encouraging for machine learning for developers as it is less complex as compared to C++ and Java. It's also a very portable language as it is used on platforms including Linux, Windows, Mac OS, and UNIX. It features interactive, interpreted, modular, dynamic, portable and high level, which make it more unique than Java.

Also, Python is a multi-paradigm programming[⑤] supporting object-oriented,

---

① functionality [ˌfʌŋkʃəˈnælətɪ] n. 功能，功能性

② autopilot [ˈɔːtəʊpaɪlət] n. 自动驾驶仪

③ gratify [ˈɡrætɪfaɪ] vt. 使高兴，使满意

④ versatility [ˌvɜːsəˈtɪlɪtɪ] n. 通用性，多用途

⑤ multi-paradigm programming: 多范式编程

procedural and functional styles of programming. Python supports neural networks and development of NLP solutions thanks to its simple function library and more so ideal structure.

Advantages：

（1）Python has a rich and extensive variety of library and tools.

（2）Supports algorithm testing without having to implement them.

（3）Python supporting object-oriented design increases a programmer's productivity.

（4）Compared to Java and C++，Python is faster in development.

Drawbacks：

（1）Developers accustomed to[1] using Python face difficulty in adjusting to completely different syntax when they try using other languages for AI programming.

（2）Unlike C++ and Java，Python works with the help of an interpreter which makes compilation and execution slower in AI development.

（3）Not suitable for mobile computing[2]. For AI meant for mobile applications，Python is unsuitable[3] due to its weak language for mobile computing.

2．C++

C++ is the fastest computer language. Its speed is appreciated for AI programming projects that are time sensitive. It provides faster execution and has less response time which is applied in search engines[4] and development of computer games. In addition，C++ allows extensive use of algorithms and is efficient in using statistical AI techniques. Another important factor is that C++ supports reuse of programs in development due to inheritance and data-hiding[5]，thus efficient in time and cost saving[6].

C++ is appropriate for machine learning and neural network.

Advantages：

（1）Good for finding solutions for complex AI problems.

（2）Rich in library functions and programming tools collection.

（3）C++ is a multi-paradigm programming that supports object-oriented principles，thus useful in achieving organized data.

Drawbacks：

---

[1] accustomed to：习惯于……

[2] mobile computing：移动计算

[3] unsuitable [ˌʌnˈsjuːtəbl] adj.不合适的，不适宜的

[4] search engine：搜索引擎

[5] data-hiding：数据隐藏

[6] data-hiding：节省成本，节省费用

（1）Poor in multitasking；C++ is suitable only for implementing core or the base of specific systems or algorithms.

（2）It follows the bottom-up approach[①], which is highly complex. It is hard for a newbie at using it to write AI programs.

3. Java

Java is also a multi-paradigm language that follows object-oriented principles and the principle of Written Once Read/Run Anywhere （WORA）[②]. It is an AI programming language that can run on any platform that supports it without the need for recompilation.

Java is one of the most commonly used languages and not just in AI development. It derives a major part of its syntax from C and C++ in addition to its lesser tools. Java is not only appropriate for NLP and search algorithms but also for neural networks.

Advantages：

（1）Very portable；it is easy to implement on different platforms because of Virtual Machine Technology[③].

（2）Unlike C++，Java is simple to use and even debug.

（3）It has an automatic memory manager which eases the work of the developer.

Disadvantages：

（1）Java is，however，slower than C++. It has less speed in execution and more response time.

（2）Though highly portable，on older platforms，Java would require dramatic changes on software and hardware to facilitate.

（3）Java is also a generally immature programming AI language as there are still some developments ongoing.

4. LISP

LISP is another language used for artificial intelligence development. It is a family of computer programming language and is the second oldest programming language after Fortran. LISP has developed over time to become strong and dynamic language in coding.

Some consider LISP as the best AI programming language due to the favour of

---

① bottom-up approach：自底向上法
② WORA：一次写成，到处可用
③ Virtual Machine Technology：虚拟机技术

liberty it offers to the developers. LISP is used in AI because of its flexibility for fast in prototyping and experimentation[①] which in turn facilitate LISP to grow to a standard AI language. For instance, LISP has a unique macro[②] system which facilitates exploration and implementation of different levels of intellectual intelligence.

LISP, unlike most AI programming languages, is more efficient in solving specific as it adapts to the needs of the solutions a developer is writing. It is highly suitable in inductive[③] logic projects and machine learning.

Advantages:

(1) Fast and efficient in coding as it is supported by compilers instead of interpreters.

(2) Automatic memory manager was invented for LISP, therefore, it has a garbage collection[④].

(3) LISP offers specific control over systems resulting to their maximum use.

Drawbacks:

(1) Few developers are well acquainted with LISP programming.

(2) Being a vintage programming language for artificial intelligence, LISP requires configuration of new software and hardware to accommodate its use.

5. Prolog

Prolog is also one of the oldest programming languages, also suitable for the development of programming AI. Like LISP, it is also a primary computer language for artificial intelligence. It has mechanisms that facilitate flexible frameworks developers enjoy working with. It is a rule-based and declarative language as it contains facts and rules that dictate its artificial intelligence coding language.

Prolog supports basic mechanisms such as pattern matching, tree-based data structuring, and automatic backtracking essential for AI programming. Other than its extensive use in AI projects, Prolog is also used for creation of medical systems.

Advantages:

(1) Prolog has a built-in list handling essential in representing tree-based data structures.

(2) It is efficient for fast prototyping for AI programs to be released modules frequently.

---

① experimentation [ɪkˌsperɪmen'teɪʃn] *n*.实验,试验
② macro ['mækrəʊ] *n*.宏
③ inductive [ɪn'dʌktɪv] *adj*.归纳法的,归纳的
④ garbage collection:垃圾回收

(3) It allows database creation simultaneous with running of the program.

Drawbacks:

Prolog has not been fully standardized in that some features differ in implementation, making the work of the developer cumbersome①.

6. Conclusion

When it comes to keeping up with technology, every individual, business person and organization do not want to be left behind. The emergence of AI technology is bringing changes that will permeate② the core of our lives, therefore understanding and using AI technology would be the best strategy right now.

参考译文

计算机软件

计算机软件由程序组成,可以让计算机执行特定的任务。它与只能机械地执行设定任务的物理构件(硬件)相对。这个术语包括应用程序(如能够提高用户工作效率的字处理器)、系统软件(如操作系统,它带有硬件接口,以便为用户界面和应用程序提供必需的服务)和中间件(管理与适应分布系统)。

1. 术语

术语"软件"是一个指令序列的程序源,它按照冯诺依曼机制制定指令流,不应该把它与配置件和流件混淆。配置件和流件都是用来配置资源(通过配置件实现结构化"编程")和制定数据流(使用流件实现数据流程编程)的程序源,是重配置计算机系统的反冯·诺依曼机制范例。

2. 与计算机硬件的关系

计算机软件是与计算机硬件相对的称谓,硬件包括物理连接和存储与执行(或运行)软件所需的设备。在计算机中,软件装入 RAM 并在中央处理器中执行。最基本的软件可以由特定处理器的机器语言组成。机器语言由一组表示处理器指令(目标代码)的二进制值组成,这些目标代码可以改变计算机的状态。软件是有序的指令序列,以特定序列用

---

① cumbersome [ˈkʌmbəsəm] adj. 麻烦的;累赘的

② permeate [ˈpɜːmɪeɪt] v. 渗入,渗透

于改变计算机硬件的状态。它通常用高级语言编写,对人来说比机器语言更便于理解且更有效(更接近自然语言)。高级语言可以编译或解释成机器语言目标代码。软件也可以用汇编语言编写,汇编语言本质上是用自然语言字母表示的机器语言助记形式。汇编语言必须通过编译器编译为目标代码。

在计算机科学和软件工程中,所有的计算机程序都是计算机软件。把不同的指令序列读到设备的内存以控制技术这一概念是由查尔斯·巴贝奇提出的,这成为他的差分机的一部分。

3.类型

实际的计算机系统把软件分为三大类:系统软件、编程软件和应用软件,尽管其差别是武断的,且通常是混淆的。

3.1 系统软件

系统软件帮助运行计算机硬件和计算机系统。它包括操作系统、设备驱动程序、诊断工具、服务程序、窗口系统、实用程序等。系统软件的目的是把应用程序员与所用的复杂计算机的细节尽可能隔离开来,尤其是与内存和其他硬件、附件(如通信设备、打印机、阅读设备、显示器、键盘等)隔开。

3.2 编程软件

编程软件通常提供帮助程序员用不同的编程语言更方便地编写计算机程序和软件的工具。这些工具包括文本编辑器、编译器、解释程序、连接程序、调试程序等。集成开发环境把这些工具合并为一个软件包,程序员不用给编译、解释、调试、跟踪等操作输入多个命令,因为 IDE 通常有高级的图形用户界面或 GUI。

3.3 应用软件

应用软件允许终端用户实现一个或多个(与计算机无关的)特定任务。典型的应用包括工业自动控制、商业软件、教育软件、医学软件、数据库和计算机游戏。商业大概是应用软件的最大用户,但几乎人类活动的每个领域现在都在使用某种应用软件。它用于各种各样的自动操作。

4.三层

用户看待事情的方法往往与程序员不同。使用现代化普通计算机(与嵌入式计算机、模拟计算机、超级计算机等不同)的人往往认为执行各种操作的软件有三个层次:平台软件、应用软件和用户软件。

4.1　平台软件

平台软件包括固件、设备驱动程序、操作系统以及有代表性的图形用户界面。总体上说，图形用户界面让用户与计算机及外设（相关设备）交互。平台软件通常与计算机捆绑提供，用户可能没有意识到它的存在或者不知道他们可以选择其他平台软件。

4.2　应用软件

应用软件或应用就是大多数人认为的软件。典型的例子包括办公套件和视频游戏。应用软件通常与计算机硬件分开购买。有时应用软件也与计算机捆绑，但这不能改变它们作为独立应用软件而运行的事实。应用软件几乎总是独立于操作系统的程序，尽管它们通常为特定的平台而制作。大部分用户把编译程序、数据库和其他"系统软件"当作应用软件。

4.3　用户软件

用户软件定制多个系统以便满足用户的特定需求。用户软件包括电子表格模板、字处理程序的宏、科学仿真及用于图形和动画的脚本。甚至电子邮件过滤器也是用户软件的一种。用户自己建立用户软件，且通常忽视它的重要性。由于用户编写软件根据其适应性被整合到所购买的应用软件包中，因而许多用户不知道所购买软件的包的差别，也不知道合作伙伴在里面加了什么。

5. 运行

计算机软件必须被"装载"到计算机的存储器（如硬盘、内存或 RAM）中。一旦软件被装入，计算机就可以执行该软件。计算机通过执行程序来运行。这包括从应用软件提取指令、经过系统软件发给最终以机器代码接收指令的硬件。每个指令都使计算机执行一个操作——移动数据、执行计算或改变指令的控制流。

数据移动通常是数据从内存中的一个位置向另一位置移动。有时数据也在内存和寄存器之间移动，寄存器可以实现在 CPU 中高速访问数据。移动数据——特别是移动大量的数据——是花费成本的。所以，有时使用"指针"来代替数据。计算包括简单的运算，如增加一个可变数据元素的值。更复杂的计算也许涉及许多运算和数据元素。

指令可以被连续地、有条件地或循环地执行。连续指令是一个接一个执行的操作。条件指令是根据某些数据的值执行不同的指令集合。在某些语言中，叫作"if"语句。循环指令是根据某些数值并反复地执行。这有时叫作一个"循环"。通常，一个指令可以调用另一个在其他程序或模块中定义的指令集合。当使用多个处理器时，指令可以同步执行。

这种软件运行方式的一个简单例子是用户从一个菜单中选择一个菜单项（如"Copy"）后所发生的一切。在这种情况下，条件指令被执行以便从内存中驻留的文本区域的数据中复制一个文本到叫作"剪切板"的一个临时存储区域。如果另一菜单项（如

"Paste")被选择,软件可以执行该指令,把剪切板数据区域中的文本复制到内存中同一文本或不同文本的特定位置。

根据应用软件,甚至以上这个例子也可以变得复杂。软件工程就是努力管理软件运行的复杂性。对于运行大的或功能强的计算机系统的软件而言,尤其如此。

按照运行软件分为以下几种:可运行的计算机程序、源代码或脚本、配置程序。

6. 软件的可靠性和质量

软件可靠性考虑与软件建立和运行相关的错误、故障及失效。

软件质量非常重要,尤其是像 Microsoft 的 Office,Microsoft Windows 和 Linux 的商业和系统软件。如果软件出现故障(出错),它可以删除一个人的工作,使计算机崩溃和做出其他意想不到的事情。故障和错误被称为"bug(漏洞)",这是 alpha 和 beta 测试过程中经常发现的。软件通常也是一个所谓的软件老化的受害者,源于看不见的错误组合而产生的渐进的性能下降。

通过软件测试可以发现和消除(调试)许多错误。然而,软件测试很少——如果有的话——消除所有的错误;有些程序员说,"每个程序至少有一个错误"。在软件开发的瀑布方法中,通常使用独立的测试团队,但在较新的方法中,统称为敏捷软件开发,开发者经常亲自做所有的测试,并定期向用户/客户展示该软件以获得反馈。软件可以通过单元测试,回归测试等方法进行测试。因为要测试的代码量可能相当大,可以手工完成,或最常见的,自动进行测试。例如,NASA 具有极为严格的许多操作系统和通信功能的软件测试程序。许多基于 NASA 的操作通过命令程序交互,相互识别。这使很多在 NASA 工作的人能够检查和评估系统的整体功能。包含命令软件的程序使硬件工程和系统操作能够更容易地共同发挥其功能。

Unit 5

录音

Text A

Knowledge Acquisition, Knowledge Representation and Knowledge-Based System

1. Knowledge Acquisition

Knowledge acquisition typically refers to the process of acquiring, processing, understanding, and recalling information through one of a number of methods. This is often a field of study closely tied to cognition, memory, and the way in which human beings are able to understand the world around them. While no single theory has been thoroughly proven or universally accepted, many theories regarding the acquisition of knowledge contain similarities that can be considered basic aspects of the process. Knowledge acquisition typically details how people experience new information, how that information is stored in the brain, and how that information can be recalled for later use.

One of the primary components of knowledge acquisition is the supposition that people are born without knowledge, and that it is gained during a person's lifetime. This is often utilized in tandem with the idea of a person as a tabula rasa or 'blank slate.' Some approaches to knowledge acquisition have been built upon the idea that people have a predisposition toward knowledge or are born with certain values or knowledge already in place. The 'blank slate' approach regards humans as essentially empty of knowledge upon birth, and that new information is acquired and utilized throughout a person's life.

Knowledge acquisition typically begins with the process of receiving or acquiring new information. This is usually done through visual, aural, and tactile signals that a person receives through his or her senses. When a person first sees a dog, for example, he or she is receiving the information about what a dog looks like. Knowledge is acquired that indicates a dog generally has four legs, is covered with fur, and has a tail.

Once information is received, knowledge acquisition typically continues through encoding and understanding that information. This encoding process allows a person to build a cognitive model, sometimes called a schema, for a piece of information. The schema for a dog, continuing the above example, incorporates the received information to build an overall sense of what constitutes 'dogness'. When a person sees another animal, such as a kangaroo, he or she processes the new information, sees that it does not fit the schema of a dog, and then creates a new model for that new knowledge.

Knowledge acquisition then continues with the ability to effectively recall and alter stored information. When someone sees a dog again, he or she is able to recognize it as a dog by recalling the schema for 'dog' and seeing that it fits into that model. This can create cognitive dissonance when someone encounters an object that exists within a certain schema, but which does not match certain aspects of that model.

Someone seeing a hairless dog for the first time, for example, may initially not fully recognize it as a dog and has to modify his or her schema for 'dog' with the newly acquired knowledge that dogs can be hairless. This entire process of knowledge acquisition usually continues throughout a person's life. It may be most intense, however, during the early years of life as someone is rapidly creating and altering schemata based on millions of different pieces of information.

2. Knowledge Representation

The field of knowledge representation involves considering artificial intelligence and how it presents some sort of knowledge, usually regarding a closed system. IT professionals and others may monitor and evaluate an artificial intelligence system to get a better idea of its simulation of human knowledge, or its role in presenting the data about focus input.

Scientists from MIT's AI Lab talk about knowledge representation as "a set of ontological commitments——a fragmented theory of intelligent reasoning" and "a simulation of a medium of human expression". Some call knowledge representation a 'surrogate' for some form of human correspondence or communication regarding a system. As artificial intelligence evolves, scientists continue to look at the boundaries and parameters of knowledge representation to further define what it means and how it applies to state-of-the-art tech knowledge as AI continues to approach better and fuller models of human sentience.

3. Knowledge-Based System

A knowledge-based system (KBS) is a computer system which generates and utilizes knowledge from different sources, data and information. These systems aid in solving problems, especially complex ones, by utilizing artificial intelligence concepts. These systems are mostly used in problem-solving procedures and to support human learning, decision making and actions.

Knowledge-based systems are considered to be a major branch of artificial intelligence. They are capable of making decisions based on the knowledge residing in them, and can understand the context of the data that is being processed.

Knowledge-based systems broadly consist of an interface engine and knowledge base. The interface engine acts as the search engine, and the knowledge base acts as the knowledge repository. Learning is an essential component of knowledge-based systems and simulation of learning helps in the betterment of the systems. Knowledge-based systems can be broadly classified as CASE-based systems, intelligent tutoring systems, expert systems, hypertext manipulation systems and databases with intelligent user interface.

Compared to traditional computer-based information systems, knowledge-based systems have many advantages. They can provide efficient documentation and also handle large amounts of unstructured data in an intelligent fashion. Knowledge-based systems can aid in expert decision making and allow users to work at a higher level of expertise and promote productivity and consistency. These systems are considered very useful when expertise is unavailable, or when data needs to be stored for future usage or needs to be grouped with different expertise at a common platform, thus providing large-scale integration of knowledge. Finally, knowledge-based systems are capable of creating new knowledge by referring to the stored content.

The limitations of knowledge-based systems are the abstract nature of the concerned knowledge, acquiring and manipulating large volumes of information or data，and the limitations of cognitive and other scientific techniques.

New Words

acquisition	[ˌækwɪˈzɪʃn]	n. 获取，获得；收集
recall	[rɪˈkɔːl]	vt. 回调，再次调用
cognition	[kɒɡˈnɪʃn]	n. 认识，认知
theory	[ˈθɪərɪ]	n. 理论；原理
universally	[ˌjuːnɪˈvɜːsəlɪ]	adv. 普遍地，一般地；人人，处处
accept	[əkˈsept]	vi. 承认，同意
regard	[rɪˈɡɑːd]	vt. 认为；注视；涉及
aspect	[ˈæspekt]	n. 方面；面貌
detail	[ˈdiːteɪl]	n. 细节；详述
supposition	[ˌsʌpəˈzɪʃn]	n. 推测；猜测；假定
gain	[ɡeɪn]	v. 获得；赢得
approach	[əˈprəʊtʃ]	v. 接近，走近，靠近
		n. 方法；途径；接近
predisposition	[ˌpriːdɪspəˈzɪʃn]	n. 倾向，素质
receive	[rɪˈsiːv]	v. 收到；接到
visual	[ˈvɪʒuəl]	adj. 视觉的，看得见的
aural	[ˈɔːrəl]	adj. 耳的；听觉的
tactile	[ˈtæktaɪl]	adj. 触觉的
signal	[ˈsɪɡnəl]	n. 信号
		vt. 向……发信号
		vi. 发信号
sense	[sens]	n. 感觉
		vt. 感到；理解，领会
indicate	[ˈɪndɪkeɪt]	vt. 表明，标示，指示
encode	[ɪnˈkəʊd]	vt. 译成密码；编码
schema	[ˈskiːmə]	n. 模式；概要，计划
incorporate	[ɪnˈkɔːpəreɪt]	v. 包含；合并；混合
kangaroo	[ˌkæŋɡəˈruː]	n. 袋鼠
dissonance	[ˈdɪsənəns]	n. 不一致，不和谐
encounter	[ɪnˈkaʊntə]	vt. 遭遇；对抗
		n. 对决，冲突；相遇，碰见
representation	[ˌreprɪzenˈteɪʃn]	n. 表现；表现……的事物
involve	[ɪnˈvɒlv]	vt. 包含；使参与，牵涉
monitor	[ˈmɒnɪtə]	vt. 监督，监控；测定

		n.显示屏,显示器
simulation	[ˌsɪmjʊˈleɪʃn]	n.模仿,模拟
input	[ˈɪnpʊt]	n.输入
ontological	[ˌɒntəˈlɒdʒɪkl]	adj.存在论的,本体论的,实体论的
fragmented	[fræɡˈmentɪd]	adj.成碎片的,片断的
surrogate	[ˈsʌrəɡɪt]	n.代理,代表
	[ˈsʌrəɡeɪt]	adj.代理的
		v.代理,替代
correspondence	[ˌkɒrəˈspɒndəns]	n.一致,符合;对应
evolve	[ɪˈvɒlv]	vt.使发展;使进化;设计
boundary	[ˈbaʊndrɪ]	n.边界,分界线;范围
parameter	[pəˈræmɪtə]	n.参数
state-of-the-art	[steɪt-əv-ðɪ-ɑːt]	adj.使用最先进技术的
sentience	[ˈsenʃəns]	n.感觉性;感觉能力;知觉
generate	[ˈdʒenəreɪt]	vt.产生,造成;形成
aid	[eɪd]	vt.辅助,帮助;促进
		n.助手;辅助设备
concept	[ˈkɒnsept]	n.观念,概念;观点;思想
action	[ˈækʃn]	n.行动,活动;功能,作用;手段
capable	[ˈkeɪpəbl]	adj.有能力的;熟练的;胜任的
repository	[rɪˈpɒzətrɪ]	n.仓库;储藏室
betterment	[ˈbetəmənt]	n.改良,改进
database	[ˈdeɪtəbeɪs]	n.数据库
traditional	[trəˈdɪʃənl]	adj.传统的;惯例的
documentation	[ˌdɒkjʊmenˈteɪʃn]	n.文档
handle	[ˈhændl]	v.操作,操控
		n.手柄;句柄
unstructured	[ʌnˈstrʌktʃəd]	adj.非结构化的,无结构的
expertise	[ˌekspɜːˈtiːz]	n.专门知识或技能;专家的意见;专家评价
promote	[prəˈməʊt]	vt.促进,推进
productivity	[ˌprɒdʌkˈtɪvɪtɪ]	n.生产率,生产力
consistency	[kənˈsɪstənsɪ]	n.一致性;符合;前后一致
usage	[ˈjuːsɪdʒ]	n.使用,用法
large-scale	[lɑːdʒ-skeɪl]	adj.大规模的,大范围的
abstract	[ˈæbstrækt]	adj.抽象的,理论上的

✑Phrases

knowledge acquisition	知识收集,知识获取
in tandem with	同……串联,同……合作
tabula rasa	白板(拉丁语)
blank slate	空白板
been built upon	建立在
cognitive model	认知模型
a piece of	一块;一片;一件
be able to	能做到……的;可以……的,能够……的
base on	基于,建立在……上
knowledge representation	知识表达,知识表现
intelligent reasoning	智能推理
human sentience	人类感觉
problem-solving procedure	问题解决过程
knowledge-based system	基于知识的系统
interface engine	接口引擎
search engine	搜索引擎
knowledge repository	知识仓库
intelligent tutoring system	智能辅导系统
hypertext manipulation system	超文本操作系统
computer-based information system	基于计算机的信息系统
unstructured data	非结构化数据

✑Abbreviations

MIT（Massachusetts Institute of Technology）	麻省理工学院
KBS（Knowledge-Based System）	知识库系统
CASE（Computer Aided Software Engineering）	计算机辅助软件工程

✑Exercises

【Ex. 1】 根据课文内容回答问题。

1. What does knowledge acquisition typically refer to?

2. What is one of the primary components of knowledge acquisition?

3. What does knowledge acquisition typically begin with? How is it done?

4. What does knowledge acquisition typically do once information is received?

5. What does the field of knowledge representation involve?

6. What do scientists from MIT's AI Lab talk about knowledge representation as?

7. What is a knowledge-based system（KBS）?

8. What is knowledge-based systems are considered to be? What are they capable of?

9. What do knowledge-based systems broadly consist of? What do they act as respectively?

10. What are the limitations of knowledge-based systems?

【Ex.2】 把下列单词或词组中英互译。

1. intelligent reasoning _____   1. _____

2. hypertext manipulation system _____   2. _____

3. knowledge representation _____   3. _____

4. search engine _____   4. _____

5. cognition _____   5. _____

6. n.数据库 _____   6. _____

7. 基于知识的系统 _____   7. _____

8. vt.译成密码；编码 _____   8. _____

9. n.参数 _____   9. _____

10. 接口引擎 _____   10. _____

【Ex.3】 短文翻译。

Knowledge Engineer

A knowledge engineer is a professional engaged in the science of building advanced logic into computer systems in order to try to simulate human decision-making and high-level cognitive tasks. A knowledge engineer supplies some or all of the 'knowledge' that is eventually built into the technology.

One principle in knowledge engineering is the transfer principle. This method involves transferring human logic and knowledge into a technology. Over time，this principle has given way to a more popular model principle，which involves the simulation of human knowledge rather than its direct transfer from human to machine.

As a specific kind of artificial intelligence project，knowledge engineering relies on a few key factors. One is a large enough repository of data to work from. Another is a complex system of algorithms that can simulate human decision-making on critical problems. Knowledge engineering is useful in decision support software projects，geographical information systems，and other new technologies that analyze data toward some higher socio-cognitive goal.

【Ex.4】　将下列词填入适当的位置(每个词只用一次)。

bases	representation	decisions	analyze	limited
solutions	rely	readable	access	knowledge

Knowledge Base

A knowledge base is a database used for knowledge sharing and management.

It promotes the collection, organization and retrieval of knowledge. Many __1__ bases are structured around artificial intelligence. They not only store data but also find __2__ for further problems using data from previous experience stored as part of the knowledge base.

Knowledge management systems depend on data management technologies ranging from relational databases to data warehouses. Some knowledge __3__ are little more than indexed encyclopedic information; others are interactive and behave/respond according to the input prompted from the user.

A knowledge base is not merely a space for data storage, but can be an artificial intelligence tool for delivering intelligent decisions. Various knowledge __4__ techniques, including frames and scripts, represent knowledge. The services offered are explanation, reasoning and intelligent decision support.

Knowledge-based computer-aided systems engineering (KB-CASE) tools assist designers by providing suggestions and solutions, thereby helping to investigate the results of design __5__. The knowledge base analysis and design allows users to frame knowledge bases, from which informative decisions are made.

The two major types of knowledge bases are human readable and machine readable.

(1) Human readable knowledge bases enable people to __6__ and use the knowledge. They store help documents, manuals, troubleshooting information and frequently answered questions. They can be interactive and lead users to solutions to problems they have, but __7__ on the user providing information to guide the process.

(2) Machine readable knowledge bases store knowledge, but only in system __8__ forms. Solutions are offered based upon automated deductive reasoning and are not so interactive as this relies on query systems that have software that can respond to the knowledge base to narrow down a solution. This means that machine readable knowledge base information shared to other machines is usually linear and is __9__ in interactivity, unlike the human interaction which is query based.

Knowledge management (KM) contains a range of strategies used in an organization to create, represent, __10__, distribute and enable the adoption of experiences. It focuses on competitive advantages and the improved performance of organizations. Work script is a well known knowledge management database.

Text B

What's a Knowledge Base and Why You Need It

1. What Is a Knowledge Base?

A knowledge base is a self-serve online library of information about a product, service, department, or topic.

The data in your knowledge base can be from anywhere, but usually comes from several contributors who are well versed on the subject—enough to give you all the details. Subjects range from the ins and outs of your HR or Legal department to how a new product, hardware, or software works. The knowledge base can include FAQs, troubleshooting guides, and any other nitty gritty details you may want or need to know.

2. How Can a KB Make a Difference for Coworkers & Customers?

When you have a knowledge base in place, you should combine it with a program of knowledge management. Knowledge management enables you to create, curate, share, utilize and manage knowledge across your whole company and across industries. With a strong knowledge base and knowledge management, you'll find your organization is more nimble and able to deliver faster service. You'll also be able to improve self-service, give greater access to more articles, and offer regular updates through that knowledge management system.

Here's another big point: by providing the opportunity for people to leave comments, you can give coworkers and customers the opportunity to help solve each other's problems. And make all involved feel a true sense of community.

Here are a few other ways a knowledge base can make a difference: more consistent service, higher resolution and lower costs.

2.1 More Consistent Service

Everybody in your organization will speak from the same playbook. No confusion. No missteps. Everyone—customers and sales folk—is happy.

2.2　Higher Resolution Rates at First Contact

With a good knowledge base, one that's organized well, there's no putting customers on hold, no transfers between agents, no "we'll call you back in a minute". Answers are right at customers' fingertips. And when they have additional questions, others in the community are right there to help. It's an easy-to-use, self-serve way to resolve issues fast.

2.3　Lower Training Costs

A knowledge base, supported by a strong knowledge management program, ensures new hires are trained with the latest information and get consistent guidance. That translates to a better work environment and lower costs. And who doesn't want that?

3. Does My Team Really Have to Have a Knowledge Base?

Look, no one's forcing you. But once you put a good knowledge base in place, backed by a plan for knowledge management, customers and employees find answers themselves. So you can focus on really important aspects of your job, rather than answering everyone's questions. Plus, there are dozens of other reasons to organize a good knowledge base. Here are just a few:

(1) It puts everything people need to know in one place and, well, it's organized.

(2) Your company looks smart, up-to-date, and professional.

(3) You standardize answers instead of offering multiple responses from different sources.

(4) You get a feedback loop and the opportunity to engage with people who matter.

(5) It's flexible. You can use it for IT tickets, for any department (even for concert tickets).

Organizations use knowledge bases for a lot of reasons. And they're finding more uses virtually every day. The way you use a knowledge base depends, of course, on what your organization does and who it serves. But here are a few of the ways knowledge bases are proving to be invaluable.

(1) IT: It simplifies everything from troubleshooting to training/onboarding and general how-to and support questions.

(2) HR: Again, great for everything from training/onboarding to distributing company policies and pay schedules.

(3) Legal: Helps with contract and other approval processes, policies, trademarks and

registrations.

Many knowledge bases are structured around artificial intelligence. Others are just indexed encyclopedias. Some are interactive and respond to input from users.

Human-readable knowledge bases are the ones people can access for documents, manuals, troubleshooting information and frequently answered questions. There are also machine-readable knowledge bases. They store stuff in system-readable forms. Solutions are based on what we call automated deductive reasoning. When a user enters a query, software helps narrow down a solution.

4. Seven Critical Tips for Building Your Knowledge Base

Besides instituting a really strong knowledge management system, here are tips on maintaining a knowledge base of your very own.

4.1 Determine That You Need One

Start by asking yourself how much time you'd save if employees didn't have to answer the same questions over and over again. Then look at your customer satisfaction and productivity goals; if your organization could do better, a knowledge base is a great place to start.

4.2 Get Your Content Together

Stuff is everywhere. Collect FAQs and answers from any department that interacts with customers. And we mean any department. The guys who build trade shows? The people in IT? Yes. And yes. But don't forget the people who make stuff— if your company makes stuff. They all can and should contribute to your knowledge base. And they should be part of the knowledge management process that maintains it.

4.3 Customize Pages. Stay Consistent

Create a style guide, so that all of the information you pour into your knowledge base looks and sounds the same. Yep, we're talking the same font, type size— formatting—all of that. This covers the entire visual presentation. From images to colors and everything in between. Look sharp. Be sharp.

4.4 Find Your Voice and Stick with It

Figure out how your company or organization talks. Whether you're polished or funky or have kind of a laid back vibe, use that in the presentation of your knowledge base. The people in marketing can help. It's a 'brand' thing.

4.5 Get the Right Tools to Manage It

We know how it is. You start out with a bunch of pdf files on a server, and your knowledge base grows from there. But once you're done, make sure you have the right tools for hosting and managing your knowledge base. Everything from the frequency of content changes to how customers typically interact with your information—all of that should be part of your decision.

4.6 Make It Easy. And Keep It That Way

Once it's up and running, remember that your knowledge base is a self-serve operation. You'll need to make sure your knowledge base is easy to navigate. And easy to use. Allow contributors to use speed templates to upload data. Use labels/search terms to categorize information so articles are easy to find. Organize the content to fit your audience and company, and then stay on top of it like a control freak's sock drawer.

4.7 Keep It Relevant. And Up To Date

This is where the creation of your knowledge base flows into the ongoing task of knowledge management. Implement a system of analytics, so you understand how many people are using content. Allow users to leave feedback and ratings on content. Make sure your management and marketing people are looking it over. Avoid delays and bottlenecks by identifying multiple people to approve content. Set up gatekeepers, then give them big red flags to wave when information needs to be dropped, added or changed.

5. Put Your Knowledge to Work for You. Now

Today, customers, employees—virtually everyone—wants and expects easy access to information. That data is locked away in the files, databases and minds inside your organization. Organizing it into a knowledge base opens the door to improved customer service, greater productivity, increased collaboration, and a lot less time spent answering the same questions.

✎ New Words

self-serve	['selfsɜːv]	adj. 自我服务的,自助的
online	['ɒn'laɪn]	adj. 在线的;联网的;联机的
topic	['tɒpɪk]	n. 主题;话题,论题

contributor	[kən'trɪbjʊtə]	n.贡献者;捐助者;投稿者
versed	[vɜːst]	adj.精通的,熟练的
troubleshooting	['trʌblʃuːtɪŋ]	n.发现并修理故障
combine	[kəm'baɪn]	v.组合,使结合
curate	['kjʊəreɪt]	v.管理
nimble	['nɪmbl]	adj.灵活的;敏捷的
deliver	[dɪ'lɪvə]	vt.交付,递送
regular	['regjʊlə]	adv.定期地;经常地
opportunity	[ˌɒpə'tjuːnɪtɪ]	n.机会
consistent	[kən'sɪstənt]	adj.一致的;连续的;坚持的
resolution	[ˌrezə'luːʃn]	n.解决;坚决;分辨率
playbook	['pleɪbʊk]	n.剧本
confusion	[kən'fjuːʒn]	n.混乱;混淆;困惑
misstep	[ˌmɪs'step]	n.错误;失策;失足
folk	[fəʊk]	n.人们;各位;大伙儿
contact	['kɒntækt]	n.接触
		vt.使接触;与……联系
	[kən'tækt]	vi.联系,接触
community	[kə'mjuːnɪtɪ]	n.社区,社团
easy-to-use	['iːzɪ-tʊ-juːz]	adj.易用的,好用的
guidance	['gaɪdns]	n.指导,引导
smart	[smɑːt]	adj.聪明的;敏捷的
standardize	['stændədaɪz]	vt.使标准化
invaluable	[ɪn'væljʊəbl]	adj.非常宝贵的;无法估计的;无价的
simplify	['sɪmplɪfaɪ]	vt.简化;使简易
trademark	['treɪdmɑːk]	n.(注册)商标
registration	[ˌredʒɪ'streɪʃn]	n.登记,注册
encyclopedia	[ɪnˌsaɪklə'piːdɪə]	n.百科全书
human-readable	['hjuːmən-'riːdəbl]	adj.人可读的
manual	['mænjʊəl]	n.手册;指南
		adj.用手的;手制的,手工的
stuff	[stʌf]	n.材料,原料,资料
deductive	[dɪ'dʌktɪv]	adj.推论的,演绎的
machine-readable	[mə'ʃiːn-'riːdəbl]	adj.机器可读的
system-readable	['sɪstəm-'riːdəbl]	adj.系统可读的
institute	['ɪnstɪtjuːt]	vt.建立;制定
satisfaction	[ˌsætɪs'fækʃn]	n.满足,满意
font	[fɒnt]	n.字体

polish	['pɒlɪʃ]	n.优美,优雅,精良
navigate	['nævɪgeɪt]	v.导航
upload	[ˌʌp'ləʊd]	vt.上传,上载
categorize	['kætəgəraɪz]	vt.把……归类,把……分门别类
creation	[krɪ'eɪʃn]	n.制造,创造
rating	['reɪtɪŋ]	n.等级;评估,评价
bottleneck	['bɒtlnek]	n.瓶颈
gatekeeper	['geɪtkiːpə]	n.看门人
collaboration	[kəˌlæbə'reɪʃn]	n.合作,协作

✑ Phrases

troubleshooting guide	故障排除指南
nitty gritty	本质;实质;基本事实
knowledge management	知识管理
leave comment	留下评论
training cost	培训费用
narrow down	(使)变窄,(使)减少,(使)缩小
over and over again	一再地;来回来去;再三再四
pour into	不断地涌进;注入;倾注
type size	字号
figure out	弄明白;解决;想出
make sure	确保
search term	搜索词

✑ Abbreviations

HR（Human Resource）	人力资源
FAQ（Frequently Asked Questions）	常见问题
IT（Information Technology）	信息技术

✑ Exercises

【Ex.5】 根据课文内容回答问题。

1. What is a knowledge base?
2. What can the knowledge base include?
3. What does knowledge management enable you to do?
4. What are the other ways mentioned in the passage that a knowledge base can make

a difference?

5. What is a knowledge base supported by? What does it ensure?

6. What are the other reasons to organize a good knowledge base?

7. What are a few of the ways knowledge bases are proving to be invaluable?

8. What are human-readable knowledge bases?

9. How do you determine that you need a knowledge base?

10. What is the last tip on maintaining a knowledge base of your own?

Reading

Automated Planning and Scheduling

Automated planning[①] and scheduling[②], sometimes denoted as simply AI Planning, is a branch of artificial intelligence that concerns the realization[③] of strategies or action sequences. Unlike classical control and classification problems, the solutions are complex and must be discovered and optimized in multidimensional[④] space. Planning is also related to decision theory[⑤].

In known environments with available models, planning can be done offline. Solutions can be found and evaluated prior to execution. In dynamically unknown environments, the strategy often needs to be revised online. Models and policies must be adapted. Solutions usually resort to iterative trial and error[⑥] processes commonly seen in artificial intelligence. These include dynamic programming, reinforcement learning and combinatorial optimization[⑦]. Languages used to describe planning and scheduling are often called action languages.

1. Overview

Given a description of the possible initial states[⑧] of the world, a description of the desired goals, and a description of a set of possible actions, the planning problem

---

① automated planning：自动规划

② scheduling ['ʃedjuːlɪŋ] n.调度；行程安排，排时间表

③ realization [ˌriːəlaɪ'zeɪʃn] n.认识，领会；实现

④ multidimensional [ˌmʌltɪdaɪ'menʃənl] adj.多维的

⑤ decision theory：决策论

⑥ trial and error：反复试验，试错法

⑦ combinatorial optimization：组合最优化

⑧ initial state：起始状态，初态

is to synthesize① a plan that is guaranteed（when applied to any of the initial states）to generate a state which contains the desired goals（such a state is called a goal state）.

The difficulty of planning is dependent on the simplifying assumptions② employed. Several classes of planning problems can be identified depending on the properties the problems have in several dimensions.

（1）Are the actions deterministic or nondeterministic③? For nondeterministic actions，are the associated probabilities available?

（2）Are the state variables discrete④ or continuous⑤? If they are discrete，do they have only a finite number of possible values?

（3）Can the current state be observed unambiguously⑥? There can be full observability⑦ and partial observability.

（4）How many initial states are there，finite or arbitrarily many?

（5）Do actions have a duration⑧?

（6）Can several actions be taken concurrently，or is only one action possible at a time?

（7）Is the objective of a plan to reach a designated goal state，or to maximize a reward function⑨?

（8）Is there only one agent or are there several agents? Are the agents cooperative or selfish? Do all of the agents construct their own plans separately，or are the plans constructed centrally for all agents?

The simplest possible planning problem，known as the Classical Planning Problem⑩，is determined by：

• a unique known initial state，
• durationless actions，
• deterministic actions，
• which can be taken only one at a time，
• and a single agent.

Since the initial state is known unambiguously，and all actions are deterministic，

---

① synthesize ['sɪnθəsaɪz] v.合成；综合
② assumption [ə'sʌmpʃn] n.假定，假设
③ nondeterministic ['nɒndɪtɜːmɪ'nɪstɪk] adj.不确定的
④ discrete [dɪ'skriːt] adj.离散的
⑤ continuous [kən'tɪnjʊəs] adj.连续的
⑥ unambiguously [ˌʌnæm'bɪgjʊəslɪ] adv.明白地，不含糊地
⑦ observability [əb'zɜːvəbɪlɪtɪ] adj.可观察性，可观测性
⑧ duration [djʊ'reɪʃn] n.持续的时间，期间
⑨ reward function：回报函数
⑩ Classical Planning Problem：经典规划问题

the state of the world after any sequence of actions can be accurately predicted, and the question of observability is irrelevant[①] for classical planning.

Further, plans can be defined as sequences of actions, because it is always known in advance which actions will be needed.

With nondeterministic actions or other events outside the control of the agent, the possible executions form a tree, and plans have to determine the appropriate actions for every node of the tree.

Discrete-time Markov decision processes (MDP) are planning problems with:

- durationless actions,
- nondeterministic actions with probabilities,
- full observability,
- maximization of a reward function,
- and a single agent.

When full observability is replaced by partial observability, planning corresponds to partially observable Markov decision process (POMDP[②]).

If there are more than one agent, we have multi-agent planning, which is closely related to game theory[③].

2. Domain Independent Planning

In AI planning, planners typically input a domain model as well as the specific problem to be solved specified by the initial state and goal, in contrast to those in which there is no input domain specified. Such planners are called 'Domain Independent' to emphasize the fact that they can solve planning problems from a wide range of domains. Typical examples of domains are block stacking, logistics, workflow management, and robot task planning. Hence a single domain independent planner can be used to solve planning problems in all these various domains. On the other hand, a route planner is typical of a domain specific planner.

3. Planning Domain Modeling Languages

The most commonly used languages for representing planning domains and specific planning problems, such as STRIPS[④] and PDDL[⑤] for Classical Planning, are

---

① irrelevant [ɪˈreləvənt] *adj*. 不相干的，不恰当
② POMDP：部分可观察马尔可夫决策过程
③ game theory：博弈论，对策论
④ STRIPS (Stanford Research Institute Problem Solver)：斯坦福研究所问题求解系统
⑤ PDDL(Planning Domain Definition Language)：规划域定义语言

based on state variables. Each possible state of the world is an assignment of values to the state variables，and actions determine how the values of the state variables change when that action is taken. Since a set of state variables induce a state space that has a size that is exponential in the set，planning，similarly to many other computational problems，suffers from the curse of dimensionality① and the combinatorial explosion.

An alternative language for describing planning problems is that of hierarchical task networks，in which a set of tasks is given，and each task can be either realized by a primitive action or decomposed into② a set of other tasks. This does not necessarily involve state variables，although in more realistic applications state variables simplify the description of task networks.

4. Algorithms for planning

4.1　Classical planning

(1) Forward chaining③ state space search④，possibly enhanced with heuristics.

(2) Backward chaining⑤ search，possibly enhanced by the use of state constraints (see STRIPS，graphplan⑥).

(3) Partial-order planning⑦.

4.2　Reduction to other problems

(1) Reduction to the propositional⑧ satisfiability problem (satplan).

(2) Reduction to model checking—both are essentially problems of traversing state spaces⑨，and the classical planning problem corresponds to a subclass of model checking problems.

4.3　Temporal planning

Temporal planning⑩ can be solved with methods similar to classical planning. The

---

① dimensionality [dɪˌmenʃəˈnælɪtɪ] n. 维度；度数
② decomposed into：分解成
③ forward chaining：正向链接，前向链接
④ state space search：状态空间搜索
⑤ backward chaining：反向链接，前后链接
⑥ graphplan：图规划
⑦ partial-order planning：偏序规划
⑧ propositional [prɑpəˈzɪʃənəl] adj. 命题的；建议的，提议的
⑨ traversing state space：遍历状态空间
⑩ temporal planning：时态规划

main difference is, because of the possibility of several temporally overlapping[①] actions with a duration being taken concurrently, that the definition of a state has to include information about the current absolute time and how far the execution of each active action has proceeded. Further, in planning with rational or real time, the state space may be infinite, unlike in classical planning or planning with integer time. Temporal planning is closely related to scheduling problems. Temporal planning can also be understood in terms of timed automata[②].

4.4 Probabilistic planning

Probabilistic planning[③] can be solved with iterative methods such as value iteration and policy iteration, when the state space is sufficiently small. With partial observability, probabilistic planning is similarly solved with iterative methods, but using a representation of the value functions defined for the space of beliefs instead of states.

4.5 Preference-based planning

In preference-based planning[④], the objective is not only to produce a plan but also to satisfy user-specified preferences. It is different from the more common reward-based planning[⑤]. Preferences don't necessarily have a precise numerical value[⑥].

4.6 Conditional planning[⑦]

Deterministic planning[⑧] was introduced with the STRIPS planning system, which is a hierarchical planner. Action names are ordered in a sequence and this is a plan for the robot. Hierarchical planning can be compared with an automatic generated behavior tree. The disadvantage is that a normal behavior tree is not so expressive like a computer program. That means the notation of a behavior graph contains action commands, but no loops or if-then-statements. Conditional planning overcomes the bottleneck[⑨] and introduces an elaborated[⑩] notation which is similar to a control flow,

---

① overlap [ˌəʊvəˈlæp] n.重叠,搭接 v.部分重叠
② automata [ɔːˈtɒmətə] n.自动操作,自动控制
③ probabilistic planning：概率规划
④ preference-based planning：基于偏好的规划
⑤ reward-based planning：基于奖励的计划
⑥ precise numerical value：精确数值
⑦ conditional planning：条件规划
⑧ deterministic planning：确定性规划
⑨ bottleneck [ˈbɒtlnek] n.瓶颈
⑩ elaborated [iˈlæbəreɪtɪd] v.详细制定；详尽阐述

known from other programming languages like Pascal. It is very similar to program synthesis[①], which means a planner generates source code which can be executed by an interpreter.

参考译文

知识获取、知识表示和基于知识的系统

1. 知识获取

知识获取通常是指通过某种方法获取、处理、理解和回调信息的过程。这通常是一个与认知、记忆以及人类能够理解周围世界的方式密切相关的研究领域。虽然没有一种理论能得到彻底证实或普遍接受，但许多关于知识获取的理论都包含可以被视为该过程基本方面的相似之处。知识获取特别详细地描述人们如何体验新信息，如何将信息存储在大脑中以及如何调用这些信息供以后使用。

知识获取的主要组成部分之一是假设人们出生时没有知识，而且知识是在其一生中获得的。这通常与人作为白板或"空白板"的想法一起使用。一些知识的获取方法是基于人们具有知识倾向或天生具有某些价值或知识的观念。"空白板"方法认为人类在出生时基本上没有知识，并且人的一生都在获取和利用新信息。

知识获取通常从接收或获取新信息的过程开始。人们通常用其感官所接收的视觉、听觉和触觉信号来实现这一过程。例如，当一个人第一次看到狗时，他或她正在接收有关狗的样子的信息。获得的知识表明狗通常有四条腿，全身长毛并且有尾巴。

一旦接收到信息，通常通过编码和理解该信息来继续获取知识。该编码过程允许人们为一条信息构建认知模型（有时称为模式）。继续上面的例子，狗的模式结合了所接收的信息，以构建对"狗"的整体感觉。当一个人看到另一只动物（例如袋鼠）时，他或她处理新信息，看到它不适合狗的模式，然后为新知识创建一个新模型。

知识获取后仍然具有有效地调用和改变存储信息的能力。当有人再次看到一只狗时，他或她能够通过回忆"狗"的模式，发现其适合该模型因而将其识别为狗。当某人遇到某个模式中存在的对象但与该模型的某些方面不匹配时，这可能会产生认知失调。

例如，有人第一次看到无毛犬，可能最初并不完全认识到它是一只狗，并且必须用新获得的知识来修改他或她的"狗"模式，即狗可以无毛。整个知识获取过程通常在整个人的生命中持续进行。然而，在人生的早期阶段，这个过程可能非常强烈。因为人一直在根据数百万条不同的信息迅速创建和改变模式。

---

① synthesis ['sɪnθəsɪs] *n*.综合；综合体；综合推理

2．知识表示

知识表示领域涉及考虑人工智能及其如何分类的知识,通常是关于封闭系统。IT专业人员和其他人员可以监控和评估人工智能系统,以更好地了解其对人类知识的模拟或其在表现焦点输入数据方面的作用。

来自麻省理工学院人工智能实验室的科学家将知识表示称为"一组本体论承诺——一种智能推理的分散理论"和"人类表达媒介的模拟"。有些人将知识表示称为某种形式的人类沟通系统或系统的通信的"代理人"。随着人工智能的发展,由于人工智能将更好地、更充分地接近人类感知模型,科学家们不断研究知识表示的边界和参数,以进一步明确它的含义以及它如何应用于最先进的技术知识。

3．基于知识的系统

基于知识的系统(KBS)是一种计算机系统,它生成和利用来自不同来源的数据和信息的知识。这些系统通过利用人工智能概念帮助解决问题,尤其是复杂问题。这些系统主要用于解决问题的程序,并支持人类学习、决策和行动。

基于知识的系统被认为是人工智能的主要分支。它们可根据驻留其中的知识做出决策,并且能够理解正在处理的数据的情景。

基于知识的系统通常由接口引擎和知识库组成。接口引擎充当搜索引擎,知识库充当知识仓库。学习是基于知识系统的重要组成部分,学习模拟有助于改善系统。基于知识的系统可大致分为基于CASE的系统、智能培训系统、专家系统、超文本操作系统和具有智能用户界面的数据库。

与传统的基于计算机的信息系统相比,基于知识的系统具有许多优点。它们可以提供有效的文档,并以智能方式处理大量非结构化数据。基于知识的系统可以辅助专家决策,并允许用户以更高的专业水平工作,提高生产力和保持一致性。当专业知识不可用时,或者需要存储数据以供将来使用,或者需要在公共平台上与不同小组的专业人士协作时,这些系统就很有用,它们能够提供大规模的知识集成。最后,基于知识的系统能够通过已经存储的内容来创建新知识。

基于知识的系统的局限性在于:相关知识具有抽象性,要获取和操纵大量的信息或数据,以及受到认知和其他科学技术的限制。

Unit 6

录音

Text A

Expert System

Expert systems are computer applications that combine computer equipment, software, and specialized information to imitate expert human reasoning and advice. As a branch of artificial intelligence, expert systems provide discipline-specific advice and explanation to their users. While artificial intelligence is a broad field covering many aspects of computer-generated thought, expert systems are more narrowly focused. Typically, expert systems function best with specific activities or problems and a discrete database of digitized facts, rules, cases, and models. Expert systems are used widely in commercial and industrial settings, including medicine, finance, manufacturing, and sales.

As a software program, the expert system integrates a searching and sorting program with a knowledge database. The specific searching and sorting program for an expert system is known as the inference engine. The inference engine contains all the systematic processing rules and logic associated with the problem or task at hand. Mathematical probabilities often serve as the basis for many expert systems. The knowledge database stores necessary factual, procedural, and experiential information representing expert knowledge. Through a procedure known as knowledge transfer, expertise, or those skills and knowledge that sustain a much better than average performance, passes from human expert to knowledge engineer. The knowledge engineer actually creates and structures the knowledge database by completing certain logical, physical, and psychosocial tasks. For this reason, expert systems are often referred to as knowledge-based information systems. By widely distributing human expertise through expert systems, businesses can realize benefits in consistency, accuracy, and reliability in problem-solving activities.

1. Building a Knowledge Base

The basic role of an expert system is to replicate a human expert and replace him or her in a problem-solving activity. In order for this to happen, key information must be transferred from a human expert into the knowledge database and, when appropriate, the inference engine. Two different types of knowledge emerge from the human expert: facts and procedural or heuristic information. Facts encompass the definitively known data and the defined variables that comprise any given activity. Procedures capture the if-then logic the expert would use in any given activity. Through a formal knowledge acquisition process that includes identification, conceptualization, formalization, implementation and testing, expert databases are developed. Interviews, transactional tracking, observation and case study are common means of extracting information from a human expert. Using programmatic and physical integration of logic, data, and choice, expert systems integrate the examination and interpretation of data input with specific rules of behavior and facts to arrive at a recommended outcome.

2. Applying Expertise: The Inference Engine

When an expert system must choose which piece of information is an appropriate answer to the specific problem at hand, uncertainty is intrinsic; thus, uncertainty is an underlying consideration in the overall conceptualization, development, and use of an expert system. One popular treatment of uncertainty uses fuzzy logic. Fuzzy logic divides the simple yes-no decision into a scale of probability. This extension of probability criteria allows the expert system to accommodate highly complex problems and activities in an attempt to more closely model human expert assistance and interaction. Probabilities of uncertainty vary from system to system based on the kind of information being stored and its intended uses.

In its diagnostic role, an expert system offers to solve a problem by analyzing yes or no with the likelihood of correctly identifying a cause of a problem or disturbance.

By inferring difficulties from past observations, the expert system identifies possible problems while offering possible advice and/or solutions. Diagnostic systems typically infer causes of problems. Applications include medicine, manufacturing, service, and a multitude of narrowly focused problem areas. As an aid to human problem solving, the diagnostic system or program assists by relying on past evidence and problems. By inferring descriptions from observations rather than problems, the expert system takes an interpretive rather than diagnostic role. Interpretive systems explain observations by inferring their meaning based on previous descriptions of situations. The probability of uncertainty is quantified as the likelihood of being an accurate representation. In a predictive role, the expert system forecasts future events and activities based on past information. Probabilities of uncertainty are emphasized as chances or the likelihood of being right. Finally, in an instructive role, the expert system teaches and evaluates the successful transfer of education information back to the user. By explanation of its decision-making process, supplemental materials and systematic testing, the instructive system accounts for uncertainty by measuring the likelihood that knowledge transfer was achieved. Regardless of the role of an expert system or how it deals with uncertainty, its anatomy is still similar. The inference engine forms the heart of the expert system. The knowledge base serves as the brain of the expert system.

The inference engine chums through countless potential paths and possibilities based on some combination of rules, cases, models, or theories. Some rules, such as predicate logic, mimic human reasoning and offer various mathematical arguments to any query. A decision tree or branching steps and actions synthesize probability with rules and information to arrive at a recommendation. Probabilities mirror the human expert's own experience with an activity or problem. Other models or cases structure some systematic movement through a problem-solving exercise in different ways. Case-based reasoning uses specific incidents or models of behavior to simulate human reasoning. Other inference engines are based on semantic networks (associated nodes and links of association), scripts (preprogrammed activities and responses), objects (self-contained variables and rule sets), and frames (more-specialized objects allowing inheritance). In all cases, the inference engine guides the processing steps and expert information together in a systematic way.

The knowledge database provides the fuel for the inference engine. The knowledge database is composed of facts, records, rules, books, and countless other resources and materials. These materials are the absolute values and documented evidence associated with the database structure. If-then procedures and pertinent rules are an important part of the knowledge database. Imitating human reasoning, rules or heuristics use logic to record expert processing steps and requirements. Logic, facts, and past experience are woven together to make an expert database. As a result of knowledge transfer, significant experiences, skills, and facts fuse together in a representation of expertise. This expert database, or knowledge-based information system, is the foil for the inference engine. As such, the knowledge database must be accurately and reliably conceived, planned, and realized for optimum performance. Additionally, the knowledge database must be validated and confirmed as accurate and reliable. Expert databases containing inaccurate information or procedural steps that result in bad advice are ineffective and potentially destructive to the operation of a business. When, however, the inference engine and knowledge database synchronize correctly, businesses may realize gains in productivity and decreases in costs.

3. Benefits and Costs

Expert systems capture scarce expert knowledge and render it archival. This is an advantage when losing the expert would be a significant loss to the organization. Distributing the expert knowledge enhances employee productivity by offering necessary assistance to make the best decision. Improvements in reliability and quality frequently appear when expert systems distribute expert advice, opinion, and explanation on demand. Expert systems are capable of handling enormously complex tasks and activities as well as an extremely rich knowledge-database structure and content. As such, they are well suited to model human activities and problems. Expert systems can reduce production downtime and, as a result, increase output and quality. Additionally, expert systems facilitate the transfer of expertise to remote locations using digital communications. In specific situations, ongoing use of an expert system may be cheaper and more consistent than the services of a human expert.

The costs of expert systems vary considerably and often include post-development costs such as training and maintenance. Prices for the software development itself range from the low thousands of dollars for a very simple system to millions for a major undertaking. For large companies and complex activities, sufficiently powerful computer hardware must be available, and frequently programming must be done to integrate the new expert system with existing information systems and process controls. Additionally, depending on the application, the knowledge database must be updated frequently to maintain relevance and timeliness. Increased costs may also appear with the identification and employment of a human expert or a series of experts. Retaining an expert involves the potentially expensive task of transferring expertise to a digital format. Depending on the expert's ability to conceive and digitally represent knowledge, this process may be lengthy. Even after such efforts some expert systems fail to recover their costs because of poor design or inadequate knowledge modeling. Expert systems suffer, as well, from the systematic integration of preexisting human biases and ignorance into their original programming.

Using an expert shell—a kind of off-the-shelf computer program for building an expert application—is one way to reduce the costs of obtaining an expert system. The expert shell simplifies the expert system by providing preprogrammed modules and a ready-to-use inference engine structure. A number of companies provide expert shells that support business and industrial operations, including those conducted in Internet environment.

✎ New Words

imitate	['ɪmɪteɪt]	vt.模仿,效仿
advice	[əd'vaɪs]	n.劝告,忠告;建议
explanation	[,eksplə'neɪʃn]	n.解释;说明
discrete	[dɪ'skriːt]	adj.分离的,不关联的
digitized	['dɪdʒɪtaɪzd]	v.数字化
engine	['endʒɪn]	n.引擎,发动机
mathematical	[,mæθə'mætɪkl]	adj.数学的;精确的
probability	[,prɒbə'bɪlɪtɪ]	n.概率;可能性,或然性
factual	['fæktʃʊəl]	adj.事实的,真实的
experiential	[ɪk,spɪərɪ'enʃl]	adj.经验的,经验上的,根据经验的
sustain	[sə'steɪn]	vt.维持;支撑,支持
psychosocial	[saɪkəʊ'səʊʃəl]	adj.社会心理的
distribute	[dɪ'strɪbjuːt]	vt.分配,散布;散发,分发
role	[rəʊl]	n.角色;作用;地位
replicate	['replɪkeɪt]	vt.复制
	['replɪkɪt]	adj.复制的
		n.复制品
emerge	[ɪ'mɜːdʒ]	vi.出现,浮现,涌现
encompass	[ɪn'kʌmpəs]	vt.围绕,包围;包含
definitively	[dɪ'fɪnɪtɪvlɪ]	adv.决定性地,最后地
formal	['fɔːml]	adj.规则的,正规的;形式的
conceptualization	[kən'septjʊəlaɪ'zeɪʃən]	n.化为概念,概念化
formalization	[,fɔːməlaɪ'zeɪʃn]	n.形式化;规则化
implementation	[,ɪmplɪmen'teɪʃn]	n.执行,履行;落实
transactional	[træn'zækʃənəl]	adj.业务的,交易的
extract	['ɪkstrækt]	vt.提取;选取;获得
programmatic	[,prəʊgrə'mætɪk]	adj.按计划的,程序的

examination	[ɪɡ͵zæmɪˈneɪʃn]	n.检查
interpretation	[ɪn͵tɜːprɪˈteɪʃn]	n.解释，说明
recommend	[͵rekəˈmend]	v.推荐；建议
appropriate	[əˈprəʊprɪət]	adj.适当的，恰当的，合适的
uncertainty	[ʌnˈsɜːtntɪ]	n.无把握，不确定；不可靠
intrinsic	[ɪnˈtrɪnsɪk]	adj.固有的，内在的，本质的
consideration	[kən͵sɪdəˈreɪʃn]	n.考虑，考察；照顾，关心
development	[dɪˈveləpmənt]	n.发展，进化
treatment	[ˈtriːtmənt]	n.处理；待遇，对待
fuzzy	[ˈfʌzɪ]	adj.模糊的
accommodate	[əˈkɒmədeɪt]	vt.容纳；使适应
assistance	[əˈsɪstəns]	n.帮助，援助；辅助设备
interaction	[͵ɪntərˈækʃn]	n.合作；互相影响；互动
intended	[ɪnˈtendɪd]	adj.有意的，预期的
diagnostic	[͵daɪəɡˈnɒstɪk]	adj.诊断的，判断的；特征的
likelihood	[ˈlaɪklɪhʊd]	n.可能，可能性；[数]似然，似真
disturbance	[dɪˈstɜːbəns]	n.打扰，困扰
difficulty	[ˈdɪfɪkəltɪ]	n.难度；困难，麻烦
multitude	[ˈmʌltɪtjuːd]	n.大量，许多
evidence	[ˈevɪdəns]	n.证据；迹象
		vt.显示；表明；证实
interpretive	[ɪnˈtɜːprɪtɪv]	adj.作为说明的，解释的
emphasize	[ˈemfəsaɪz]	vt.强调，着重；使突出
chance	[tʃɑːns]	n.机会，机遇
instructive	[ɪnˈstrʌktɪv]	adj.有益的；指导性的
		adv.有益地；有启发性地
		n.教育；指导性
anatomy	[əˈnætəmɪ]	n.分解，分析
chum	[tʃʌm]	vi.结交，成为好朋友
		n.密友
countless	[ˈkaʊntlɪs]	adj.无数的，多得数不清的
potential	[pəˈtenʃl]	adj.潜在的，有可能的
combination	[͵kɒmbɪˈneɪʃn]	n.结合；联合体
mimic	[ˈmɪmɪk]	vt.模仿，模拟
		adj.模仿的
argument	[ˈɑːɡjʊmənt]	n.论据，理由，论点
synthesize	[ˈsɪnθəsaɪz]	v.综合，合成
mirror	[ˈmɪrə]	vt.反映，反射

exercise	['eksəsaɪz]	vi.训练,练习
		n.练习,训练;运用
incident	['ɪnsɪdənt]	n.事件
associate	[ə'səuʃɪeɪt]	vt.(使)发生联系;(使)联合
link	[lɪŋk]	vt.链接,连接
		n.链接;关联
response	[rɪ'spɒns]	n.响应,反应;回答,答复
object	['ɒbdʒɪkt]	n.对象;物体;目标
frame	[freɪm]	n.框架
inheritance	[ɪn'herɪtəns]	n.继承;遗传
systematic	[ˌsɪstə'mætɪk]	adj.系统的,规则的;有步骤的
record	['rekɔːd]	n.记录
absolute	['æbsəluːt]	adj.绝对的,完全的
		n.绝对;绝对事物
pertinent	['pɜːtɪnənt]	adj.有关的,相干的;恰当的
significant	[sɪg'nɪfɪkənt]	adj.重要的;显著的;有重大意义的
fuse	[fjuːz]	vi.融入;融合
		vt.使融合;使融化
reliably	[rɪ'laɪəblɪ]	adv.可靠地,确实地
conceive	[kən'siːv]	v.构思;想象,设想
inaccurate	[ɪn'ækjurɪt]	adj.不精确的;不准确;不正确的
destructive	[dɪ'strʌktɪv]	adj.破坏性的;毁灭性的;有害的
synchronize	['sɪŋkrənaɪz]	vt.使同步;使同时
		vi.同时发生;共同行动
capture	['kæptʃə]	vt.&n.捕获,捕捉
scarce	[skeəs]	adj.罕见的
enhance	[ɪn'hɑːns]	vt.提高,增加;加强
enormously	[ɪ'nɔːməslɪ]	adv.巨大地,庞大地
extremely	[ɪk'striːmlɪ]	adv.极端地;非常,很
downtime	['dauntaɪm]	n.停工期
considerably	[kən'sɪdərəblɪ]	adv.相当,非常,颇
sufficiently	[sə'fɪʃntlɪ]	adv.足够地,充分地;十分,相当
timeliness	['taɪmlɪnɪs]	n.及时性
expensive	[ɪk'spensɪv]	adj.昂贵的,花钱多的
lengthy	['leŋθɪ]	adj.长的,漫长的
inadequate	[ɪn'ædɪkwɪt]	adj.不适当的;不足胜任的;不充足的
preexist	['priːɪg'zɪst]	v.先前存在的,预先存在的
bias	['baɪəs]	n.偏见;倾向

ignorance	[ˈɪɡnərəns]	*n*. 无知，愚昧
off-the-shelf	[ɒf-ðə-ʃelf]	*adj*. 现成的，买来不用改就可用的
ready-to-use	[ˈredɪ-tʊ-juːz]	*adj*. 即用的，随时可用的
conduct	[kənˈdʌkt]	*v*. 引导；实施；执行

✎ Phrases

knowledge database	知识数据库
inference engine	推理机，推理引擎
at hand	在手边，在附近；即将来临
serve as	充当，担任
knowledge engineer	知识工程师
knowledge-based information system	知识库信息系统
case study	案例研究，个案研究；范例分析
fuzzy logic	模糊逻辑
diagnostic system	诊断系统
supplemental material	辅助材料
systematic testing	系统测试
predicate logic	谓词逻辑
mimic human reasoning	模仿人类推理
mathematical argument	数学论证
case-based reasoning	实例推理（法），基于案例推理
semantic network	语义网络
self-contained variable	自包含变量
rule set	规则集
absolute value	绝对值
weave together	编织在一起
optimum performance	最佳性能
digital communication	数字通信系统
post-development cost	开发后成本
digital format	数字格式
knowledge modeling	知识模型化，知识建模
a kind of ...	……的一种

✎ Exercises

【Ex. 1】 根据课文内容回答问题。

1. What are expert systems?

2. What do expert systems function best with typically?

3. What is the basic role of an expert system?

4. What do facts encompass?

5. What does an expert system do in its diagnostic role?

6. What does the expert system do in a predictive role?

7. How does the inference engine chum through countless potential paths and possibilities?

8. What is the knowledge database composed of?

9. What are expert systems capable of?

10. What is an expert shell? How does it simplify the expert system?

【Ex.2】 把下列单词或词组中英互译。

1. anatomy	1. _____
2. capture	2. _____
3. conceptualization	3. _____
4. discrete	4. _____
5. diagnostic	5. _____
6. n.解释；说明	6. _____
7. adj.模糊的	7. _____
8. 诊断系统	8. _____
9. 知识数据库	9. _____
10. 最佳性能	10. _____

【Ex.3】 短文翻译。

Expert System

In artificial intelligence，an expert system is a computer system that emulates the decision-making ability of a human expert.

Expert systems are designed to solve complex problems by reasoning about knowledge like an expert，and not by following the procedure of a developer as is the case in conventional programming. The first expert systems were created in the 1970s and then proliferated in the 1980s. Expert systems were among the first truly successful forms of AI software.

An expert system has a unique structure. It is different from traditional programs. It is divided into two parts，one fixed，independent of the expert system：the inference engine，and one variable：the knowledge base. To run an expert system，the engine reasons about the knowledge base like a human. In the 80's a third part appeared：a dialog interface to communicate with users. This ability to conduct a conversation with users was later called 'conversational'.

【Ex.4】 将下列词填入适当的位置（每个词只用一次）。

stored	inference	reasoning	unreliable	conclusion
determine	expert	engine	fuzzy	inconsistent

Inference Engine

Complete inference engine can be divided into following functional elements:

(1) Control system—determines the order of testing the knowledge base rules.

(2) Rules interpreter—defines a Boolean (true, not true uncertainty factor) applications rules.

(3) Explanation mechanism—justifies user the process of ___1___ and generates report.

A very important element of the expert system is also the inference engine. Knowledge of the science must always be ___2___ in the knowledge base, in formalized form, understandable to the inference engine. Using symbols you can easily ___3___ how to handle the system to solve and analyze the correctness of the knowledge base. Since the inference engine is separated from the knowledge base it can be used in skeletal ___4___ systems.

Reasoning comes with certain pattern, which allows tasks to be based on the veracity of premises. It is an attempt to determine the truth of the hypothesis targeted by inference ___5___. It is based on the assumption that between sentences there is an objective ___6___ ratio, or the ratio of probabilities. Inference requires the ability to make decisions based on their knowledge.

The inference is divided into reliable and ___7___. An example of a reliable inference is deductive reasoning. A special variant thereof is syllogistic reasoning from two premises. Syllogism model contains major and minor premise, which show the application. The inference unreliable evidence does not warrant the truthfulness of the ___8___. The direction of this inference is considered to be ___9___ with the direction of logical consequence. Such inference is unreliable.

There are three basic types of reasoning: back (regressive), forward (progressive), and mixed. There are also methods of inference that use the uncertain knowledge. An example of this technique is called ___10___ logic. The most commonly used and most important methods of inference in expert systems are forward and backward chaining.

Text B

Why We Might Need an 'Interpreter' to Build Our Trust in AI

A team of researchers is working to build trust between humans and artificial intelligence (AI) by creating an 'interpreter' that can explain how an AI arrived at the answer to a specific question.

In an age of self-driving cars and autonomous drones, AI is becoming a bigger part of our lives. It's also getting increasingly savvy. Today, AI can recognize text, distinguish people by their faces, and even identify physical objects, to some degree. But even the best AI systems still get things wrong much of the time.

That poses a big problem, says Kate Saenko, assistant professor of computer science at Boston University.

"If an AI tool makes mistakes, human users quickly learn to discount it, and eventually stop using it altogether," she says. "I think that humans by nature are not likely to just accept things that a machine tells them."

"A further complication," she adds, "is that as AI becomes more powerful, the algorithms that drive it have become increasingly opaque to human users. Information goes into one end of a computational 'black box', and an answer comes out the other side—yet the set of rules and reasoning used to find that answer are obscured."

Saenko is working to change that relationship. Her research seeks to uncover new ways of getting inside the 'mind' of AI, creating a translation tool that explains its decision-making process to human users.

On the surface, that goal may sound trivial. Who cares how a computer comes to an answer, as long as it's right? Getting feedback on why an AI device makes a particular decision, however, may ultimately help improve its accuracy by giving opportunities for humans to offer tiny course corrections, Saenko says.

In the process, it could increase the trust that humans put into a machine, making it a better collaborator on complex jobs. Achieving that sort of openness in today's AI, though, may not be so simple.

1. New Types of AI

It hasn't always been hard to look inside the mind of AI. In the past, many artificial intelligence systems, like facial recognition, used rules and guidelines that programmers identified ahead of time—rules for defining skin color, for what shapes make up a nose, for defining light and shadow. All those user-created concepts had to be hard-coded into AI from the start, giving it a framework to do its job.

This method makes it fairly simple to figure out how a machine came to its conclusion: just identify which preprogrammed rules it used to get there. It also fundamentally limits the abilities of AI. Real life is vastly complex after all and even the best human programmers can't come up with every possible rule that a computer might use to make sense of the world.

"It's very hard for us to anticipate all possible ways a dog might look in any image anywhere in the world, for example," says Saenko. "If you have enough processing power and data, a better approach would be to show a computer a million pictures of dogs and let it define them itself."

In the last five years or so, that approach has become more widely used in the AI world. Instead of working with a single template, these new systems involve a more iterative approach, modeled on the way that our own nervous system works.

These new types of AI, called 'deep neural networks', employ huge numbers of interconnected functions, or nodes, arranged in a vast web. Each one is responsible for parsing a tiny amount of information and progressively builds on the work of the nodes before it.

This sort of incremental process, building bit by bit on simple data, is at the core of a deep neural network. It makes AI flexible, fast, and powerful—for some systems, it can operate with more than 95 percent accuracy. In those few cases where it's not accurate, though, deep neural networks make it extremely hard to figure out why. There are no preset coded definitions to turn to, since a neural network creates those big-picture guidelines as it goes.

"The sheer number of parameters these models can process is the reason they've been good at visual and language tasks like automated translation. It lets them soak up a lot of data," says collaborator Trevor Darrell of the University of California, Berkeley. "But because they have so many parameters, it's very difficult to directly extract and interpret structures within them."

2. Creating an AI 'Interpreter'

Saenko and Darrell are working with Zeynep Akata, a colleague at the University of Amsterdam in the Netherlands, and Kitware, an open-source software company, on ways to crack into deep neural networks and make them more easily understood.

Asking a network like this to explain itself would likely reduce its speed and efficiency, the researchers say, so they're hoping to create a sort of translation tool— a second network that acts alongside the first, interpreting its choices in real time and reporting them to a human user.

"The primary neural network is just doing its job. All of its processing is just devoted to solving its task, like finding doors or windows in an image, for example," says Saenko. "That's why we want to use a second neural network that has access to that machinery and input data, and can learn to translate all that into a textual version that humans can understand."

"This translation is important," Saenko says, "because when a deep neural network does make a mistake, it's probably because it found a pattern in the data that doesn't quite match the real world. If it's steering an autonomous car along a poorly maintained road, for instance, it might stop at a shadow, thinking it's a massive pothole."

If that happens, an 'interpreter' AI could prompt a user for more information in plain English. "I want it to be able to say, 'I stopped driving because I'm not sure if that's a pothole or a shadow, so tell me what to do here,'" she says.

"In the future, we're going to be using AI as a collaboration between humans and computers. We need to be able to communicate with it, understand its strengths, and know what it's good at, so it can help us with things we're not so good at—like sorting through a petabyte of video to identify content," Saenko says. "I see this as creating super humans. It's a collaboration between humans and AI."

✎ New Words

researcher	[rɪˈsɜːtʃə]	n. 研究员，调查者
explain	[ɪkˈspleɪn]	v. 说明，解释
autonomous	[ɔːˈtɒnəməs]	adj. 自治的，有自主权的
drone	[drəʊn]	n. 无人机
savvy	[ˈsævɪ]	n. 机智；头脑
		adj. 有见识的
distinguish	[dɪˈstɪŋgwɪʃ]	vi. 区分，辨别，分清
altogether	[ˌɔːltəˈgeðə]	adv. 全部地；完全地；总而言之
complication	[ˌkɒmplɪˈkeɪʃn]	n. 纠纷；混乱
obscure	[əbˈskjʊə]	adj. 不清楚的；隐蔽的
		vt. 使……模糊不清；掩盖，隐藏
uncover	[ʌnˈkʌvə]	vi. 发现，揭示
trivial	[ˈtrɪvɪəl]	adj. 无价值的；平常的，平凡的；不重要的
correction	[kəˈrekʃn]	n. 修改
		adj. 改正的，纠正的
collaborator	[kəˈlæbəreɪtə]	n. 协作者，合作者

openness	['əʊpənnɪs]	n. 开放，公开
conclusion	[kən'kluːʒn]	n. 断定，推论；结论，结局
preprogrammed	[priː'prəʊgræmd]	adj. 预编程序的
fundamentally	[ˌfʌndə'mentəlɪ]	adv. 基础地；根本地；从根本上
nervous	['nɜːvəs]	adj. 神经系统的
responsible	[rɪ'spɒnsəbl]	adj. 尽责的；承担责任；负有责任的
progressively	[prə'gresɪvlɪ]	adv. 日益增加地；逐步
preset	[ˌpriː'set]	vt. 预设，预先布置；事先安排
machinery	[mə'ʃiːnərɪ]	n. 机器，机械装置
textual	['tekstʃʊəl]	adj. 文本的，正文的，原文的
steer	[stɪə]	v. 驾驶；操纵，控制；引导
pothole	['pɒthəʊl]	n. 坑
strength	[streŋθ]	n. 力量；优点，长处

✎ Phrases

arrived at	到达
autonomous drone	自主无人机
physical object	物理物体
translation tool	翻译工具
decision-making process	决策程序，决策过程
facial recognition	人脸识别，面孔识别，面部识别
preprogrammed rule	预编程序规则
nervous system	神经系统
soak up	吸收，(使)充满
super human	超人

✎ Exercises

【Ex.5】 根据课文内容回答问题。

1. What is a team of researchers is working to do? And how?
2. What can AI do today?
3. What does Saenko's research seek to do?
4. What may getting feedback on why an AI device makes a particular decision do?
5. What did many artificial intelligence systems, like facial recognition, used in the past?
6. What does deep neural networks do?
7. Why is it very difficult to directly extract and interpret structures within these models?

8. What are Saenko and Darrell working with Zeynep Akata on ways to do?

9. Why is translating all input data into a textual version that humans can understand very important according to Saenko?

10. What are we going to be using AI as in the future according to Saenko?

Reading

Natural Language Processing

Natural language processing (NLP) is a branch of artificial intelligence that helps computers understand, interpret and manipulate① human language. NLP draws from many disciplines, including computer science and computational linguistics②, in its pursuit to fill the gap between human communication and computer understanding.

1. Evolution of natural language processing

While natural language processing isn't a new science, the technology is rapidly advancing thanks to an increased interest in human-to-machine communications, plus an availability of big data, powerful computing and enhanced algorithms.

As a human, you may speak and write in English, Spanish or Chinese. But a computer's native language-known as machine code or machine language-is largely incomprehensible③ to most people. At your device's lowest levels, communication occurs not with words but through millions of zeros and ones that produce logical actions.

Indeed, programmers used punch cards④ to communicate with the first computers 70 years ago. This manual and arduous⑤ process was understood by a relatively small number of people. Now you can say, "Alexa, I like this song," and a device playing music in your home will lower the volume and reply, "OK. Rating saved," in a humanlike voice. Then it adapts its algorithm to play that song-and others like it-the next time you listen to that music station.

---

① manipulate [mə'nɪpjʊleɪt] vt. 操作,处理

② linguistic [lɪŋ'gwɪstɪk] adj. 语言的,语言学的

③ incomprehensible [ɪnˌkɒmprɪ'hensəbl] adj. 难以理解的,难懂的

④ punch card: 穿孔卡片

⑤ arduous ['ɑːdjʊəs] adj. 艰巨的,难克服的

Let's take a closer look at that interaction. Your device activated when it heard you speak, understood the unspoken intent in the comment, executed an action and provided feedback in a well-formed English sentence, all in the space of about five seconds. The complete interaction was made possible by NLP, along with other AI elements such as machine learning and deep learning.

2. Why is NLP important?

2.1 Large volumes of textual data

Natural language processing helps computers communicate with humans in their own language and scales other language-related tasks. For example, NLP makes it possible for computers to read text, hear speech, interpret it, measure sentiment and determine which parts are important.

Today's machines can analyze more language-based data than humans, without fatigue[①] and in a consistent, unbiased[②] way. Considering the staggering[③] amount of unstructured data that's generated every day, from medical records to social media, automation will be critical to fully analyze text and speech data efficiently.

2.2 Structuring a highly unstructured data source

Human language is astoundingly[④] complex and diverse. We express ourselves in infinite ways, both verbally and in writing. Not only are there hundreds of languages and dialects[⑤], but within each language is a unique set of grammar and syntax rules, terms and slang[⑥]. When we write, we often misspell or abbreviate words, or omit punctuation. When we speak, we have regional accents[⑦], and we mumble[⑧], stutter[⑨] and borrow[⑩] terms from other languages.

While supervised and unsupervised learning, and specifically deep learning, are now widely used for modeling human language, there's also a need for syntactic and semantic understanding and domain expertise that are not necessarily present in these

---

① fatigue [fə'tiːg] *n.* 疲劳,疲乏

② unbiased [ʌn'baɪəst] *adj.* 无偏见的,不偏不倚的,公正的

③ staggering ['stægərɪŋ] *adj.* 难以置信的,令人震惊

④ astoundingly [ə'staʊndɪŋlɪ] *adv.* 令人震惊地

⑤ dialect ['daɪəlekt] *n.* 方言,土语;语调

⑥ slang [slæŋ] *n.* 俚语;黑话

⑦ accent ['æksent] *n.* 重音;口音;腔调

⑧ mumble ['mʌmbl] *v.* 咕哝 *n.* 含糊的话;咕哝

⑨ stutter ['stʌtə] *vt. & vi.* 结结巴巴地说 *n.* 结巴,口吃

⑩ borrow ['bɒrəʊ] *v.* 借用

machine learning approaches. NLP is important because it helps resolve ambiguity[①] in language and adds useful numeric structure to the data for many downstream applications, such as speech recognition or text analytics.

3. How does NLP work?

Natural language processing includes many different techniques for interpreting human language, ranging from statistical and machine learning methods to rules-based and algorithmic approaches. We need a broad array of approaches because the text- and voice-based data varies widely, as do the practical applications.

Basic NLP tasks include tokenization and parsing, lemmatization[②]/stemming[③], part-of-speech tagging, language detection and identification of semantic relationships. If you ever diagramed sentences in grade school, you've done these tasks manually before.

In general terms, NLP tasks break down language into shorter, elemental pieces, try to understand relationships between the pieces and explore how the pieces work together to create meaning.

These underlying tasks are often used in higher-level NLP capabilities, such as:

(1) Content categorization[④]. A linguistic-based document summary, including search and indexing, content alerts and duplication detection.

(2) Topic discovery and modeling. Accurately capture the meaning and themes in text collections, and apply advanced analytics to text, like optimization and forecasting.

(3) Contextual extraction. Automatically pull structured information from text-based sources.

(4) Sentiment analysis. Identifying the mood or subjective opinions within large amounts of text, including average sentiment and opinion mining.

(5) Speech-to-text and text-to-speech conversion. Transforming voice commands into written text, and vice versa.

(6) Document summarization[⑤]. Automatically generating synopses[⑥] of large bodies of text.

(7) Machine translation. Automatic translation of text or speech from one language to another.

---

① ambiguity [ˌæmbɪˈgjuːətɪ] *n*. 含糊；意义不明确；歧义
② lemmatization [ˌlemətaɪˈzeɪʃən] *n*. 词形还原
③ stemming [ˈstemɪŋ] *n*. 词干提取
④ content categorization：内容分类
⑤ summarization [ˌsʌməraɪˈzeɪʃən] *n*. 摘要，概要；概括
⑥ synopses [sɪˈnɑːpsiːz] *n*. 摘要，梗概；大纲

In all these cases，the overarching goal is to take raw language input and use linguistics and algorithms to transform or enrich the text in such a way that it delivers greater value.

4．NLP methods and applications

4.1　NLP and text analytics

Natural language processing goes hand in hand[①] with text analytics，which counts，groups and categorizes words to extract structure and meaning from large volumes of content. Text analytics is used to explore textual content and derive new variables from raw text that may be visualized[②]，filtered，or used as inputs to predictive models or other statistical methods.

NLP and text analytics are used together for many applications，including：

（1）Investigative[③] discovery. Identify patterns and clues in emails or written reports to help detect and solve crimes.

（2）Subject-matter expertise. Classify content into meaningful topics so you can take action and discover trends.

（3）Social media analytics. Track awareness and sentiment about specific topics and identify key influencers.

4.2　Everyday NLP examples

There are many common and practical applications of NLP in our everyday lives. Beyond conversing with virtual assistants like Alexa or Siri，here are a few more examples.

（1）Have you ever looked at the emails in your spam folder and noticed similarities[④] in the subject lines? You're seeing Bayesian spam filtering[⑤]，a statistical NLP technique that compares the words in spam to valid emails to identify junk mail[⑥].

（2）Have you ever missed a phone call and read the automatic transcript of the voicemail[⑦] in your email inbox or smartphone app? That's speech-to-text conversion，

---

① hand in hand：手拉手，密切合作；携手

② visualized [ˈvɪʒʊəˌlaɪzd] *adj*.直观的，直视的 *v*.设想，想象

③ investigative [ɪnˈvestɪɡətɪv] *adj*.调查性质的，研究的

④ similarity [ˌsɪməˈlærətɪ] *n*.类似；相像性；类似性

⑤ spam filtering：垃圾邮件过滤

⑥ junk mail：垃圾广告邮件

⑦ voicemail [ˈvɔɪsmeɪl] *n*.语音信箱

an NLP capability.

(3) Have you ever navigated a website by using its built-in search bar, or by selecting suggested topic, entity or category tags? Then you've used NLP methods for search, topic modeling, entity extraction and content categorization.

A subfield① of NLP called natural language understanding (NLU) has begun to rise in popularity because of its potential in cognitive and AI applications. NLU goes beyond the structural understanding of language to interpret intent, resolve context and word ambiguity, and even generate well-formed human language on its own. NLU algorithms must tackle② the extremely complex problem of semantic interpretation——that is, understanding the intended meaning of spoken or written language, with all the subtleties, context and inferences that we humans are able to comprehend.

The evolution of NLP toward NLU has a lot of important implications for businesses and consumers alike. Imagine the power of an algorithm that can understand the meaning and nuance③ of human language in many contexts, from medicine to law to the classroom. As the volumes of unstructured information continue to grow exponentially, we will benefit from computers' tireless④ ability to help us make sense of it all.

参考译文

专 家 系 统

专家系统是计算机应用程序,它结合了计算机设备、软件和专业信息,以模仿专家的人工推理和建议。作为人工智能的一个分支,专家系统为其用户提供特定学科的建议和解释。虽然人工智能涵盖计算机生成思想的许多领域,但专家系统的关注点更为狭隘。通常,专家系统最适合于特定的活动或问题以及数字化事实、规则、案例和模型的离散数据库。专家系统广泛用于商业和工业环境,包括医药、金融、制造和销售。

作为软件程序,专家系统将搜索和排序程序与知识数据库集成在一起。专家系统的特定搜索和排序程序称为推理引擎。推理引擎包含与手头的问题或任务相关的所有系统处理规则和逻辑。数学概率通常作为许多专家系统的基础。知识数据库存储代表专家知识的必要事实、程序和经验信息。通过被称为知识转移的程序,专业知识即那些超过平均水平的技能和知识从人类专家传递给知识工程师。知识工程师通过完成某些逻辑、物理

---

① subfield ['sʌbfiːld] n. 子域
② tackle ['tækl] vt. 着手处理
③ nuance ['njuːɑːns] n. 细微差别;细微的表情
④ tireless ['taɪəlɪs] adj. 不疲倦的,孜孜不倦的

和心理社会任务来实际地创建和构建知识数据库。因此,专家系统通常被称为基于知识的信息系统。通过专家系统广泛分发人类专业知识,利于企业在解决问题的活动中实现一致性、准确性和可靠性。

1. 建立一个知识库

专家系统的基本作用是复制人类专家并在解决问题的活动中取代他或她。为了实现这一点,必须将关键信息从人类专家转移到知识数据库中,在适当的情况下也转移给推理引擎。人类专家有两种不同类型的知识:事实和程序式或启发式信息。事实包括明确已知的数据和组成任何给定活动的已定义变量。程序捕获专家在任何给定活动中使用的if-then逻辑。通过包括识别、概念化、形式化、实施和测试的正式知识获取过程,专家数据库被开发了。访谈、业务跟踪、观察和案例研究是从人类专家中获得信息的常用手段。通过使用逻辑、数据和选择进行编程和物理集成,专家系统将数据输入的检查和解释与特定的行为规则和事实相结合,以得出推荐结论。

2. 应用专长:推理引擎

当专家系统必须选择用哪一条信息来回答手头特定问题时,不确定性是固有的;因此,不确定性是专家系统整体概念化、开发和使用时要考虑的基本因素。一种流行的处理不确定性的方法是使用模糊逻辑。模糊逻辑将简单的"是—否"决策划分为概率范围。这种概率准则的扩展让专家系统能够适应高度复杂的问题和活动,以便更接近地模拟人类专家的辅助和交互。根据存储的信息类型及其预期用途,不确定性的概率因系统而异。

在其诊断作用中,专家系统通过分析是或否以及正确识别问题或干扰原因的可能性来解决问题。通过从过去的观察中推断出困难,专家系统在提供可能的建议和/或解决方案的同时识别可能的问题。诊断系统通常会推断出问题的原因。其应用包括医药、制造、服务以及众多狭窄的问题领域。作为人类解决问题的辅助手段,诊断系统或程序通过依据过去的证据和问题来提供帮助。通过观察而不是问题来推断描述,专家系统做出解释而非诊断。解释系统通过基于先前对情况的描述来推断其意义来解释观察。不确定性的概率被量化为准确表示的可能性。在做预测时,专家系统根据过去的信息预测未来的事件和活动。强调不确定性的概率是机会或正确的可能性。最后,作为一种指导性的角色,专家系统教导并评估成功传回给用户的教育信息。通过解释其决策过程、补充材料和系统测试,指导系统通过测量知识转移的可能性来解释不确定性。除了专家系统的作用或其处理不确定性的方法,其解析结构都相似。推理引擎构成了专家系统的核心。知识库充当专家系统的大脑。

推理引擎基于规则、案例、模型或理论的某些组合,通过无数潜在的路径和可能性工作。一些规则(例如谓词逻辑),它们模仿人类推理并为任何查询提供各种数学参数。决策树或分支步骤和动作将概率与规则和信息整合以得出推荐结论。概率反映了人类专家自己对活动或问题的经验。其他模型或案例通过不同方式解决问题来构建一些系统。基于案例的推理使用特定事件或行为模型来模拟人的推理。其他推理引擎的基础是语义网络(关联节点和关联链接)、脚本(预编程活动和响应)、对象(自包含变量和规则集)和帧(允许继承的更专用对象)。在所有情况下,推理引擎引导把处理步骤和专家信息系统地整合在一起。

知识数据库为推理引擎提供燃料。知识数据库由事实、记录、规则、书籍和无数其他资源和材料组成。这些材料绝对有价值并且是与数据库结构相关的记录证据。"如果—那么"程序和相关规则是知识数据库的重要部分。模仿人类推理,规则或启发式方法使用逻辑来记录专家处理步骤和要求。逻辑、事实和过去的经验被编织在一起,形成一个专家数据库。作为知识转移的结果,重要的经验、技能和事实融合在一起,代表了专业知识。该专家数据库或基于知识的信息系统是推理引擎的载体。因此,必须准确、可靠地构思、规划和实现知识数据库,以获得最佳性能。此外,必须验证知识数据库并确认其准确可靠。包含导致错误建议的不准确信息或程序步骤的专家数据库是无效的,并且可能对企业的运营具有破坏性。但是,当推理引擎和知识数据库正确同步时,企业可以提高生产率和降低成本。

3．好处和成本

专家系统捕获稀缺的专家知识并将其归档。当失去专家对组织来说是一个重大损失时,这是一个优势。分发专家知识可给员工提供必要的帮助使他们做出最佳决策,从而提高了员工的工作效率。当专家系统按需分发专家建议、意见和解释时,就会提高可靠性和质量。专家系统能够处理极其复杂的任务和活动以及极其丰富的数据库结构和内容。因此,它们非常适合模拟人类活动和问题。专家系统可以减少生产停机时间,从而提高产量和质量。另外,专家系统利用数字通信很容易把专业知识传递给远方。在特定情况下,长期使用专家系统可能比人类专家的服务更便宜且更一致。

专家系统的成本差异很大,通常包括培训和维护等后期开发成本。软件开发本身的价格差异很大,有价值数千美元的一个非常简单的系统,也有价值数百万美元的一个重要业务的系统。对于大型公司和复杂的活动,必须提供足够强大的计算机硬件,并且必须经常进行编程以将新的专家系统与现有的信息系统和过程控制集成在一起。此外,根据应用程序的要求,必须经常更新知识库,以保持相关性和及时性。要识别和雇用人类专家或一系列专家,也会增加成本。聘用专家来完成将专业知识转换为数字格式的任务也可能很昂贵。根据专家的能力来构思和数字化表示知识,这个过程可能很漫长。即使经过这些努力,一些专家系统由于设计不良或知识建模不足而无法收回成本。由于原有程序系统中存在人类的偏见和无知,专家系统与其整合时也会受到影响。

使用框架式专家系统(一种构建专家应用程序的现成计算机程序)是降低获取专家系统成本的一种方法。专家系统通过提供预编程模块和即用型推理引擎结构简化了专家系统。许多公司提供支持商业和工业运营的框架式专家系统,可在互联网环境中操作。

Unit 7

录音

Text A

What Is Machine Learning?

Artificial intelligence and machine learning are among the most trending technologies these days. Artificial intelligence teaches computers to behave like a human, to think, and to give a response like a human, and to perform the actions like humans perform.

1. What is Machine Learning?

As the name suggests, machine learning means the machine is learning.

This is the technique through which we teach the machines about things. It is a branch of artificial intelligence and I would say it is the foundation of artificial intelligence. Here we train our machines using data. If you take a look into it, you'll see that it is something like data mining. Actually, the concept behind it is that machine learning and data mining are both data oriented. We work on data in both of situations. Actually, in data sciences or big data, we analyze the data and make the statistics out of it and we work on how we can maintain our data, how we can conclude the results and make a summary of it instead of maintaining the complete comprehensive bulk of data. But in machine learning, we teach the machines to make the decisions about things. We teach the machine with different data sets and then we check the machine for some situations and see what kind of results we get from this unknown scenario. We also use this trained model for prediction in new scenarios.

We teach the machine with our historical data, observations, and experiments. And then, we predict with the machine from these learnings and take the responses.

As I already said, machine learning is closely related to data mining and statistics.

(1) Data mining—Concerned with analytics of data.

(2) Statistics—Concerned with prediction-making/probability.

2. Why Do We Need Machine Learning?

In this era, we're using wireless communication, internet etc. Using social media, or driving cars, or anything we're doing right now, is actually generating the data at the backend. If you're surprised about how our cars are generating the data, remember that every car has a small computer inside which controls your vehicle completely, i. e., when which component needs the current, when the specific component needs to start or switch. In this way, we're generating TBs (terabytes) of data.

But this data is also important to get to the results. Let's take an example and try to understand the concept clearly. Let's suppose a person is living in a town and he goes to a shopping mall and buys something. We have many items of a single product. When he buys something, now we can generate the pattern of the things he has bought. In the same way, we can generate the selling and purchasing patterns of things of different people. Now you might be thinking about a random person who comes and buys something and then he never comes again, but we have the pattern of things there as well. With the help of this pattern we can make a decision about the things people most like and when they come to the mall again. They will see the things they want just at the entrance. This is how we attract the customer with machine learning.

3. How Do Machines Learn?

Actually, machines learn through the patterns of data. Let's start with the data sets of data. The input we give to the machine is called X and the response we get is Y. Here we've three types of learning.

1) Supervised Learning

2) Unsupervised Learning

3) Reinforcement Learning

3.1 Supervised Learning

In supervised learning, we know about the different cases (inputs) and we know the labels (output) of these cases. And here we already know about the basic truths, so here we just focus on the function (operation) because it is the main and most important thing here(See Figure 7-1).

Here we just create the function to get the output of the inputs. And we try to create the function which processes the data and try to give the accurate outputs (Y) in most of the scenarios(See Figure 7-2).

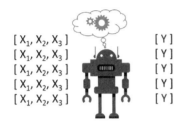

Figure 7-1   Inputs and output of supervised learning   Figure 7-2   Function of supervised learning

Because we've started with known values for our inputs, we can validate the model and make it even better.

And now we teach our machine with different data sets. Now it is the time to check it in unknown cases and generate the value.

Note: Let's suppose you've provided the machine a data set of some kind of data and now you train the model according to this data set. Now the result comes to you from this model on the basis of this knowledge set you've provided. But let's suppose if you delete an existing item in this knowledge set or you update something then you don't expect the results you get according to this new modification you've made in the data set.

3.2 Unsupervised Learning

It is quite different from supervised learning. Here we don't know about the labels (output) of different cases. And here, we train the model with patterns by finding similarities. And then these patterns become the cluster(See Figure 7-3).

Cluster = Collection of similar patterns of data

And then, this cluster is used to analyze and to process the data.

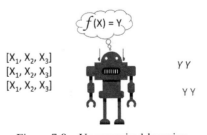

Figure 7-3   Unsupervised learning

In unsupervised learning, we really don't know if the output is right or wrong. So here in this scenario, the system recognizes the pattern and tries to calculate the results until we get the nearly right value.

3.3　Reinforcement Learning

It is like reward based learning. The example of reward based is, suppose your parents will give you a reward on the completion of a specific task. So here you know you've to complete this task and the time you need to complete it. The developer decides himself what reward he'll give on the completion of this task.

It is also feedback oriented learning. Now you're doing some tasks and on the basis of these tasks, you're getting feedback. And if the feedback is positive then it means you're doing it right and you can improve your work on your own. And if the feedback is negative then you know as well what was wrong and how to do it correctly. And feedback comes from the environment where it is working.

It makes the system more optimal than the unsupervised scenario, because here we have some clues like rewards or good feedback to make our system more efficient.

4. Steps in Machine Learning

There are some key point steps of machine learning when we start to teach the machine.

4.1　Collect data

As we already know machine learning is data oriented. We need data to teach our system for future predictions.

4.2　Prepare the input data

Now you've downloaded the data, but when you're feeding the data, you need to make sure of the particular order of the data to make it meaningful for you machine learning tool to process it; i.e. .csv file (comma separated value). This is the best format of the file to process the data because comma separated values help a lot in clustering.

4.3　Analyze data

Now, you're looking at the patterns in the data to process it in a better way. You're checking the outliers (scope & boundaries) of the data. And you are also checking the novelty (specification) of the data.

4. 4　Train model

This is the main part of the machine learning when you are developing the algorithm where you are structuring the complete system with coding to process the input and give back the output.

4. 5　Test the model

Here you're checking the values you're getting from the system whether it matches your required outcome or not.

4. 6　Deploy it in the application

Let's discuss an example of autonomous cars which don't have human intervention, which run on their own. The first step is to collect the data, and you have to collect many kinds of data. You're driving the car which runs on its owe. The car should know the road signs, it should have the knowledge of traffic signals and when people crossing the road, so that it can make the decision to stop or run in different situations. So we need a collection of images of these different situations, it is our collect data module.

Now we have to make the particular format of data (images) like. csv file where we store the path of the file, the dimensions of the file. It makes our system processing efficient. This is what we called preparing the data.

And then it makes the patterns for different traffic signals (red, green, yellow), for different sign boards of traffic and for its environment in which car or people running around. Next it decides the outlier of these objects whether it is static (stopped) or dynamic (running state). Finally it can make the decision to stop or to move in the side of another object.

These decisions are obviously dependent upon training the model, and what code we write to develop our model. This is what we say training the model and then we test it and then we deploy it in our real world applications.

5. Applications of Machine Learning

Machine learning is widely used today in our applications.

(1) You might use the Snapchat or Instagram app where you can apply the different animal's body parts on your face like ears, nose, tongue etc. These different organs places at the exact right spot in the image, this is an application of machine learning.

(2) Google is a widely used AI, ML. Google Lens is an application. If you scan

anything through Google lens then it can tell the properties and features of this specific thing.

(3) Google Maps is also using machine learning. For example，if you're watching any department store on the map，sometimes it is telling to you how much something is priced，and how expensive it is.

✎New Words

action	['ækʃn]	n.行动,活动；功能,作用
train	[treɪn]	v.训练；教育；培养
situation	[ˌsɪtjʊ'eɪʃn]	n.（人的）情况；局面,形势,处境；位置
conclude	[kən'kluːd]	vt.得出结论；推断出；决定
comprehensive	[ˌkɒmprɪ'hensɪv]	adj.广泛的；综合的
bulk	[bʌlk]	n.大块,大量；大多数,大部分
scenario	[sə'nɑːrɪəʊ]	n.设想；可能发生的情况；剧情梗概
backend	[bækend]	n.后端
random	['rændəm]	adj.任意的；随机的
		n.随意；偶然的行动
attract	[ə'trækt]	vt.吸引；引起……的好感（或兴趣）
		vi.具有吸引力；引人注意
function	['fʌŋkʃn]	n.函数
accurate	['ækjʊrət]	adj.精确的,准确的；正确无误的
validate	['vælɪdeɪt]	vt.确认；证实
knowledge	['nɒlɪdʒ]	n.知识；了解,理解
delete	[dɪ'liːt]	v.删除
update	[ˌʌp'deɪt]	vt.更新
	['ʌpdeɪt]	n.更新
gradually	['grædʒʊəlɪ]	adv.逐步地,渐渐地
horizontally	[ˌhɒrɪ'zɒntəlɪ]	adv.水平地,横地
vertical	['vɜːtɪkl]	adj.垂直的,竖立的
		n.垂直线,垂直面
approximately	[ə'prɒksɪmɪtlɪ]	adv.近似地,大约
reward	[rɪ'wɔːd]	n.奖赏；报酬；赏金；酬金
		vt.奖赏；酬谢
optimal	['ɒptɪməl]	adj.最佳的,最优的；最理想的
feed	[fiːd]	vt.馈送；向……提供
comma	['kɒmə]	n.逗号
separated	['sepəreɪtɪd]	adj.分隔的,分开的

novelty	['nɒvltɪ]	n.新奇,新奇的事物
deploy	[dɪ'plɔɪ]	v.使展开；施展；有效地利用
intervention	[ˌɪntə'venʃn]	n.介入,干涉,干预
path	[pɑːθ]	n.路径
decide	[dɪ'saɪd]	vt.决定；解决；裁决
		vi.决定；下决心
static	['stætɪk]	adj.静止的；不变的
organ	['ɔːgən]	n.器官；元件

✎ Phrases

as the name suggests	顾名思义
take a look into	看一看
data oriented	数据导向
be concerned with	涉及；与……有关
wireless communication	无线通信
knowledge set	知识集
straight line	直线
road sign	交通标志,路标
traffic signal	交通信号；红绿灯

✎ Abbreviations

| TB（TeraByte） | 兆字节 |

✎ Exercises

【Ex. 1】 根据 Text A 课文内容填空。

1. Machine learning means _____ . It is _____ through which we teach the machines about things. It is a branch of _____ .

2. We teach the machine with _____ , _____ and _____ . And then，we predict with the machine from these learnings and _____ .

3. Machine learning is closely related to _____ and _____ . Machine learning is _____ .

4. In supervised learning, we know about _____ and we know _____ .

5. Because we've started with known values for our inputs，we can _____ and _____ .

6. In unsupervised learning, we really don't know _____ .

7. Reinforcement learning is like _____ . It is also _____ .

8. There are some key point steps of machine learning when we start to teach the machine. They are _____ , _____ , _____ , _____ , _____ , and _____ .

9. _____ is the best format of the file to process the data. Because comma separated values help a lot in _____ .

10. Train model is the main part of the machine learning when you are _____ where you are structuring the complete system with coding to _____ and _____ .

【Ex.2】 把下列单词或词组中英互译。

1. knowledge set   1. _____
2. wireless communication 2. _____
3. data oriented   3. _____
4. bulk   4. _____
5. backend   5. _____
6. adj.任意的；随机的   6. _____
7. vt.更新   7. _____
8. vt.确认；证实   8. _____
9. n.设想；可能发生的情况   9. _____
10. n.函数   10. _____

【Ex.3】 短文翻译。

Decision Tree

A decision tree is a graphical representation of specific decision situations that are used when complex branching occurs in a structured decision process. A decision tree is a predictive model based on a branching series of boolean tests that use specific facts to make more generalized conclusions.

Decision trees are a popular and powerful tool used for classification and prediction purposes. Decision trees provide a convenient alternative for viewing and managing large sets of business rules，allowing them be translated in a way that allows humans to understand them and apply the rules constraints in a database so that records falling into a specific category are sure to be retrieved.

Decision trees generally consist of the following four steps：

（1）Structuring the problem as a tree by creating end nodes of the branches，which are associated with a specific path or scenario along the tree.

（2）Assigning subject probabilities to each represented event on the tree.

（3）Assigning payoffs for consequences. This could be a specific dollar amount or utility value that is associated with a particular scenario.

（4）Identifying and selecting the appropriate course（s）of action based on analyses.

【Ex.4】 将下列词填入适当的位置（每个词只用一次）。

interact	compare	doctors	search	pilot's
finance	choices	applications	respond	make

What Are the Applications of Artificial Intelligence?

There are many different applications of artificial intelligence and new uses for the technology are developed each year as artificial intelligence programs become more sophisticated. Artificial intelligence is frequently used by the military and in aviation and robotics. It is also used in the public sector in medicine， __1__ ，and business. Artificial intelligence has even made intelligent toys available for children at relatively low prices.

One of the most obvious __2__ of artificial intelligence is in creating computerized brains for robots. These programs allow robots to make choices about how to respond to stimuli，without direct input from humans. One example of this type of technology is the rovers that are deployed on the surfaces of distant planets. These machines make __3__ about how to get around obstacles because the great distance between the robot and humans makes it impractical for the robot to wait for each new instruction to arrive. Less sophisticated versions of these programs are often used in toys，creating robots with a limited ability to __4__ with their environment and with the children playing with them.

There are also applications of artificial intelligence in the field of medicine. Intelligent diagnostic programs can make observations about a patient's symptoms and __5__ that data to possible syndromes that the patient could be affected with. This technology is extremely useful because it is easy for human __6__ to overlook a condition，whereas computer programs cannot forget to take information into account.

Military organizations also find many applications for artificial intelligence. Programs are used to run simulations which can help humans __7__ important tactical decisions. Additionally，artificial intelligence is used in aviation，where it can assist in training pilots. These programs can also gather information about a __8__ abilities，which helps it learn to train each pilot more efficiently.

Many people come into contact with artificial intelligence in customer service programs. These programs，often called chat bots，are programmed to __9__ to inquiries from clients and offer support or answers to questions. Human customer service representatives are often made available if the artificial intelligence program is unable to help.

Financial institutions also find applications for artificial intelligence. Specialized programs are designed to ___10___ for patterns in financial markets in order to make investment decisions. These programs are capable of initiating trades on their own and often make decisions that are at least as financially sound as those of human brokers.

Text B

Deep Learning

Deep learning is a machine learning technique that teaches computers to do what comes naturally to humans: learn by example. Deep learning is a key technology behind driverless cars, enabling them to recognize a stop sign, or to distinguish a pedestrian from a lamppost. It is the key to voice control in consumer devices like phones, tablets, TVs, and hands-free speakers. Deep learning is getting lots of attention lately and for good reason. It's achieving results that were not possible before.

In deep learning, a computer model learns to perform classification tasks directly from images, text, or sound. Deep learning models can achieve state-of-the-art accuracy, sometimes exceeding human-level performance. Models are trained by using a large set of labeled data and neural network architectures that contain many layers.

1. How Does Deep Learning Attain Such Impressive Results?

In a word, accuracy. Deep learning achieves recognition accuracy at higher levels than ever before. This helps consumer electronics meet user expectations, and it is crucial for safety-critical applications like driverless cars. Recent advances in deep learning have improved to the point where deep learning outperforms humans in some tasks like classifying objects in images.

While deep learning was first theorized in the 1980s, there are two main reasons it has only recently become useful:

(1) Deep learning requires large amounts of labeled data. For example, driverless car development requires millions of images and thousands of hours of video.

(2) Deep learning requires substantial computing power. High-performance GPUs have a parallel architecture that is efficient for deep learning. When combined with clusters or cloud computing, this enables development teams to reduce training time

for a deep learning network from weeks to hours or less.

2. Examples of Deep Learning at Work

Deep learning applications are used in industries from automated driving to medical devices.

(1) Automated Driving: Automotive researchers are using deep learning to automatically detect objects such as stop signs and traffic lights. In addition, deep learning is used to detect pedestrians, which helps decrease accidents.

(2) Aerospace and Defense: Deep learning is used to identify objects from satellites that locate areas of interest, and identify safe or unsafe zones for troops.

(3) Medical Research: Cancer researchers are using deep learning to automatically detect cancer cells. Teams at UCLA built an advanced microscope that yields a high-dimensional data set used to train a deep learning application to accurately identify cancer cells.

(4) Industrial Automation: Deep learning is helping to improve worker safety around heavy machinery by automatically detecting when people or objects are within an unsafe distance of machines.

(5) Electronics: Deep learning is being used in automated hearing and speech translation. For example, home assistance devices that respond to your voice and know your preferences are powered by deep learning applications.

3. How Deep Learning Works

Most deep learning methods use neural network architectures, which is why deep learning models are often referred to as deep neural networks.

The term 'deep' usually refers to the number of hidden layers in the neural network. Traditional neural networks only contain $2 \sim 3$ hidden layers, while deep networks can have as many as 150.

Deep learning models are trained by using large sets of labeled data and neural network architectures that learn features directly from the data without the need for manual feature extraction.

One of the most popular types of deep neural networks is known as convolutional neural networks (CNN or ConvNet). A CNN convolves learned features with input data, and uses 2D convolutional layers, making this architecture well suited to processing 2D data, such as images.

CNNs eliminate the need for manual feature extraction, so you do not need to identify features used to classify images. The CNN works by extracting features directly from images. The relevant features are not pretrained; they are learned while the network trains on a collection of images. This automated feature extraction makes deep learning models highly accurate for computer vision tasks such as object classification.

CNNs learn to detect different features of an image using tens or hundreds of hidden layers. Every hidden layer increases the complexity of the learned image features. For example, the first hidden layer could learn how to detect edges, and the last learns how to detect more complex shapes specifically catered to the shape of the object we are trying to recognize.

4. What's the Difference Between Machine Learning and Deep Learning?

Deep learning is a specialized form of machine learning. A machine learning workflow starts with relevant features being manually extracted from images. The features are then used to create a model that categorizes the objects in the image. With a deep learning workflow, relevant features are automatically extracted from images. In addition, deep learning performs 'end-to-end learning', where a network is given raw data and a task to perform, such as classification, and it learns how to do this automatically.

Another key difference is deep learning algorithms scale with data, whereas shallow learning converges. Shallow learning refers to machine learning methods that plateau at a certain level of performance when you add more examples and training data to the network.

A key advantage of deep learning networks is that they often continue to improve as the size of your data increases.

In machine learning, you manually choose features and a classifier to sort images. With deep learning, feature extraction and modeling steps are automatic.

5. Choosing Between Machine Learning and Deep Learning

Machine learning offers a variety of techniques and models you can choose based on your application, the size of data you're processing, and the type of problem you want to solve. A successful deep learning application requires a very large amount of data (thousands of images) to train the model, as well as GPUs, or graphics processing units, to rapidly process your data.

When choosing between machine learning and deep learning, consider whether you have a high-performance GPU and lots of labeled data. If you don't have either of those things, it may make more sense to use machine learning instead of deep learning. Deep learning is generally more complex, so you'll need at least a few thousand images to get reliable results. Having a high-performance GPU means the model will take less time to analyze all those images.

6. How to Create and Train Deep Learning Models

6.1 Training from Scratch

To train a deep network from scratch, you gather a very large labeled data set and design a network architecture that will learn the features and model. This is good for new applications, or applications that will have a large number of output categories. This is a less common approach because with the large amount of data and rate of learning, these networks typically take days or weeks to train.

6.2 Transfer Learning

Most deep learning applications use the transfer learning approach, a process that involves fine-tuning a pretrained model. You start with an existing network, such as AlexNet or GoogLeNet, and feed in new data containing previously unknown classes. After making some tweaks to the network, you can now perform a new task, such as categorizing only dogs or cats instead of 1000 different objects. This also has the advantage of needing much less data (processing thousands of images, rather than millions), so computation time drops to minutes or hours.

Transfer learning requires an interface to the internals of the pre-existing network, so it can be surgically modified and enhanced for the new task. MATLAB® has tools and functions designed to help you do transfer learning.

6.3 Feature Extraction

A slightly less common, more specialized approach to deep learning is to use the network as a feature extractor. Since all the layers are tasked with learning certain features from images, we can pull these features out of the network at any time during the training process. These features can then be used as input to a machine learning model such as support vector machines (SVM).

7. Accelerating Deep Learning Models with GPUs

Training a deep learning model can take a long time, from days to weeks. Using GPU acceleration can speed up the process significantly. Using MATLAB with a GPU reduces the time required to train a network and can cut the training time for an image classification problem from days down to hours. In training deep learning models，MATLAB uses GPUs（when available）without requiring you to understand how to program GPUs explicitly.

✎New Words

naturally	[ˈnætʃərəlɪ]	adv.自然地,顺理成章地；合理地
lamppost	[ˈlæmppəʊst]	n.灯杆,路灯柱
exceed	[ɪkˈsiːd]	vt.超过,超越,胜过
attain	[əˈteɪn]	v.达到,实现；获得
crucial	[ˈkruːʃl]	adj.关键性的,极其显要的；决定性的
outperform	[ˌaʊtpəˈfɔːm]	vt.做得比……更好,胜过
substantial	[səbˈstænʃl]	adj.大量的；重大的；结实的,牢固的
automotive	[ˌɔːtəˈməʊtɪv]	adj.汽车的；自动的
automatically	[ˌɔːtəˈmætɪklɪ]	adv.自动地
detect	[dɪˈtekt]	vt.检测,发现
aerospace	[ˈeərəʊspeɪs]	n.航天
defense	[dɪˈfens]	n.国防,防卫
satellite	[ˈsætəlaɪt]	n.卫星,人造卫星
cell	[sel]	n.细胞
microscope	[ˈmaɪkrəskəʊp]	n.显微镜
unsafe	[ʌnˈseɪf]	adj.不安全的,危险的
translation	[trænsˈleɪʃn]	n.翻译
extraction	[ɪkˈstrækʃn]	n.取出,提取
convolve	[kənˈvɒlv]	v.卷积
pretrained	[priːˈtreɪnd]	adj.预训练的
shape	[ʃeɪp]	n.形状；模型 vi.使成形；形成
workflow	[ˈwɜːkfləʊ]	n.工作流程,工作流
shallow	[ˈʃæləʊ]	adj.浅的,肤浅的
converge	[kənˈvɜːdʒ]	vi.收敛
plateau	[ˈplætəʊ]	n.平稳时期,稳定水平；停滞期 v.达到平稳状态；进入停滞期
high-performance	[haɪ-pəˈfɔːməns]	adj.高性能的
fine-tune	[faɪn-tjuːn]	vt.微调,调整
tweak	[twiːk]	vt.稍稍调整(机器、系统等)
surgically	[ˈsɜːdʒɪklɪ]	adv.精确地,如外科手术般地
acceleration	[əkˌseləˈreɪʃn]	n.加速

| explicitly | [ɪk'splɪsɪtlɪ] | *adv*. 明白地,明确地 |

✍Phrases

deep learning	深度学习
driverless car	无人驾驶汽车
stop sign	停车标志
hands-free speaker	免提扬声器
labeled data	标记数据,标签化数据
safety-critical application	安全关键应用
parallel architecture	并行体系结构
cloud computing	云计算
medical device	医疗设备
high-dimensional data set	高维数据集
heavy machinery	重型机械
convolutional layer	卷积层
hidden layer	隐藏层
end-to-end learning	端到端学习
shallow learning	浅层学习
transfer learning	迁移学习
pre-existing network	现存网络
feature extraction	特征提取
feature extractor	特征提取器

✍Abbreviations

GPU（Graphics Processing Unit）	图形处理器
UCLA（University of California at Los Angeles）	加利福尼亚大学洛杉矶分校
CNN（Convolutional Neural Network）	卷积神经网络
SVM（Support Vector Machines）	支持向量机

✍Exercises

【Ex.5】 根据课文内容回答问题。

1. What is deep learning? What does a computer model do in deep learning?

2. What are the two main reasons deep learning has only recently become useful?

3. Where are deep learning application used?

4. What structure do most deep learning methods use? What does the term 'deep'

usually refer to?

5. What is one of the most popular types of deep neural networks known as?

6. What is the difference between deep learning and machine learning?

7. What is a key advantage of deep learning networks?

8. What does a successful deep learning application require?

9. What approach do most deep learning applications use? What is it?

10. What are the results of using GPU?

Reading

Some Machine Learning Algorithms

Machine learning algorithms are divided into three broad categories: supervised learning, unsupervised learning and reinforcement learning. The following are some machine learning algorithms that everyone involved in data science, machine learning and AI should know about.

1. Supervised Learning

Supervised learning is the task of inferring[①] a function from the training data. The training data consists of a set of observations[②] together with its outcome. This is used when you have labeled data sets available to train e.g. a set of medical images of human cells/organs that are labeled as malignant or benign. Supervised learning can be further subdivided into[③] regression analysis, and classification analysis.

1.1 Regression Analysis

Regression analysis is used to predict numerical values. The following are top regression algorithms.

1.1.1 Linear Regression[④]

Linear regression is a basic and commonly used type of predictive analysis[⑤]. The overall idea of regression is to examine two things: (1) does a set of predictor

---

① infer [ɪnˈfɜː] vt. 推断，推理
② observation [ˌɒbzəˈveɪʃn] n. 观察；评论
③ subdivide into：细分为
④ linear regression：线性回归
⑤ predictive analysis：预测分析

variables do a good job in predicting an outcome（dependent）variable?（2）Which variables in particular are significant predictors of the outcome variable，and in what way do they—indicated by the magnitude① and sign of the beta estimates—impact the outcome variable? These regression estimates are used to explain the relationship between one dependent variable and one or more independent variables. The methodology provides insights into the factors that have a greater influence② on the outcome，for example，the color of an automobile may not have a strong correlation to its chances of breaking down，but the make/model may have a much stronger correlation(See Figure 7-4).

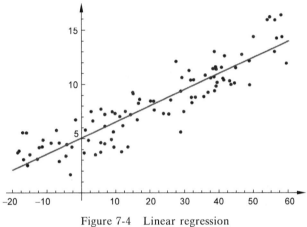

Figure 7-4　Linear regression

1. 1. 2　Polynomial Regression

Polynomial regression③ is a form of regression analysis in which the relationship between the observation and the outcome is modeled as an nth degree polynomial④. The method is more reliable when the curve is built on a large number of observations that are distributed in a curve or a series of humps，and not linear(See Figure 7-5).

1. 2　Classification Analysis

Classification analysis is a series of techniques used to predict categorical values，i. e. assign data points to categories e. g. spam email⑤ vs non-spam email，or red vs blue vs green objects. The top classification algorithms are as follows.

---

① magnitude ['mægnɪtjuːd] *n*. 重大，重要
② influence ['ɪnfluəns] *n*. 影响　*vt*. 影响；支配；对……起作用
③ polynomial regression：多项式回归
④ polynomial [ˌpɒlɪ'nəʊmɪəl] *adj*. 多项式的
⑤ spam email：垃圾邮件

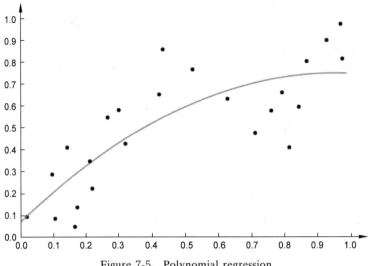

Figure 7-5   Polynomial regression

1.2.1   Logistic Regression

Logistic regression[1] is a misleading[2] name even though the name suggests regression but in reality, it is a classification technique. It is used to estimate the probability of a binary (1 or 0) response e. g. malignant or benign. It can be generalized to predict more than two categorical values, for example, is the object an animal, human, or car?

1.2.2   K-Nearest Neighbor

K-nearest neighbors[3] is a classification technique where an object is classified by a majority[4] vote. Suppose you are trying to classify the image of a flower as either sunflower or rose, and if k is chosen as 3, then 2 or all 3 of the 3 nearest classified neighbors should belong to the same flower class for the test sample[5] image to be assigned that flower class. Nearness is measured for each dimension that is used for classification, for example, color and how close the color of the test sample to the color of other pre-classified flower images. It is neighbors that the observation is assigned to the class which is most common among its K-nearest neighbors. The best choice of k depends upon the data generally. The larger value of k reduces the effect of noise on the classification number.

---

①   logistic regression：逻辑回归

②   misleading [ˌmɪsˈliːdɪŋ] adj. 误导性的；引入歧途的

③   K-nearest neighbor：K 最近邻

④   majority [məˈdʒɒrəti] n. 多数

⑤   test sample：测试样本，试样

1. 2. 3 Decision Trees

Decision trees[①] is a decision support tool that uses a tree-like model of decisions. The possible consequences[②] decision trees aim to create is a model that predicts by learning simple decision rules from the training data(See Figure 7-6).

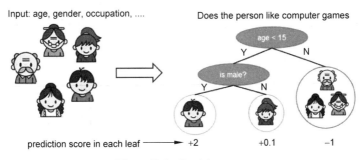

Figure 7-6   Decision trees

2. Unsupervised Learning

Unsupervised learning is a set of algorithms used to draw inferences from data sets consisting of input data without using the outcome. The most common unsupervised learning method is cluster analysis[③] which is used for exploratory[④] data analysis to find hidden patterns or groupings in data. The popular unsupervised learning algorithms are as follows.

2. 1 K-means Clustering

K-means clustering[⑤] aims to partition observations into k clusters, for instance, the item in a supermarket are clustered in categories like butter, cheese, and milk—a group dairy products. K-means algorithm does not necessarily find the most optimal configuration[⑥]. The K-means algorithm is usually run multiple times to reduce this effect(See Figure 7-7).

2. 2 Principal Component Analysis

Principal component analysis[⑦] is a technique for feature extraction when faced

---

① decision tree：决策树
② consequence ['kɒnsɪkwəns] *n*. 结果，结论，推论
③ cluster analysis：聚类分析
④ exploratory [ɪk'splɒrətri] *adj*. 探索的，考察的
⑤ K-means clustering：K-均值聚类
⑥ configuration [kən,fɪgə'reɪʃn] *n*. 构造，配置
⑦ principal component analysis：主成分分析

with too many features or variables. Say you want to predict the GDP of a country, you have many variables to consider-inflation[①], stock data for index funds[②] as well as individual stocks[③], interest rate[④], jobless claims, unemployment rate[⑤], and the list goes on. Working with too many variables is problematic for machine learning as there can be risk of overfitting[⑥], lack of suitable data for each variable, and degree of correlation of each variable on the outcome. The first principal component has the largest possible variance[⑦] that accounts for as much of the variability[⑧] in the data as possible. Each succeeding component, in turn, has the highest variance possible under the constraint that it is orthogonal to the preceding component.

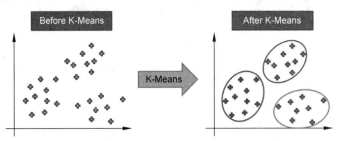

Figure 7-7   K-means clustering

3. Reinforcement Learning

Reinforcement learning is different from both supervised and unsupervised learning. The goal in supervised learning is to find the best label based on past history of labeled data, and the goal in unsupervised learning is to assign logical grouping of the data in absence[⑨] of outcomes or labels. In reinforcement learning, the goal is to reward good behavior, similar to rewarding pets for good behavior in order to reinforce that behavior.

Reinforcement learning solves the difficult problem of correlating immediate actions with the delayed outcomes they create. Like humans, reinforcement learning

---

① inflation [ɪnˈfleɪʃn] n.通货膨胀
② index fund：指数基金
③ individual stock：个股
④ interest rate：利率
⑤ unemployment rate：失业率
⑥ overfitting [ˈəʊvəˈfɪtɪŋ] n.过拟合
⑦ variance [ˈveərɪəns] n.方差
⑧ variability [ˌveərɪəˈbɪlətɪ] n.变化性，变率
⑨ absence [ˈæbsəns] n.缺乏，缺少

algorithms sometimes have to contend with delayed gratification[1] to see the outcomes of their actions or decisions made in the past, for example, the rewarding for a win in a game of chess or maximizing the points won in a game of Go with AlphaGo over many moves.

Top reinforcement learning algorithms include Q-Learning, State-Action-Reward-State-Action (SARSA[2]), Deep Q-Network (DQN), and Deep Deterministic Policy Gradient (DDPG[3]). The explanation for these algorithms gets fairly involved.

参考译文

什么是机器学习？

人工智能和机器学习是当今最流行的技术。人工智能教导计算机表现得像人一样，去思考，去做出像人一样的反应，并像人一样行事。

1. 什么是机器学习？

顾名思义，机器学习意味着机器正在学习。

这是我们教会机器有关事物的技术。它是人工智能的一个分支，我想说这是人工智能的基础。在这里，我们使用数据训练机器。如果你仔细研究一下，就会发现它类似于数据挖掘。实际上，其背后的概念是机器学习和数据挖掘都是面向数据的。我们在两种情况下都处理数据。实际上，在数据科学或大数据中，我们分析数据并从中进行统计，然后研究如何维护数据，如何得出结果并对其进行总结，而不是维护完整的全面的数据。但是在机器学习中，我们教会机器做出有关事物的决策。我们使用不同的数据集教机器，然后检查机器的某些情况，并查看从这种未知情况中得到的结果。我们还将这种训练好的模型用于新场景的预测。

我们用历史数据、观察结果和实验来教机器。然后，我们根据这些经验对机器进行预测并做出响应。

就像我已经说过的那样，机器学习与数据挖掘和统计紧密相关。

（1）数据挖掘——与数据分析有关。

（2）统计——与预测/概率有关。

---

① gratification [ˌɡrætɪfɪˈkeɪʃn] n.满足，满意
② SARSA：状态-动作-奖励-状态-动作
③ DDPG：深度确定性策略梯度

2．为什么我们需要机器学习？

在这个时代，我们正在使用无线通信、互联网等。使用社交媒体、驾驶汽车或我们现在正在做的任何事情，实际上是在后端生成数据。如果你对我们的汽车如何生成数据感到惊讶，请记住，每辆汽车内部都有一台小计算机，可以完全控制车辆，即哪个组件何时需要电流，何时需要启动或切换特定组件。通过这种方式，我们可以生成 TB（兆字节）的数据。

但是，这些数据对于获得结果也很重要。让我们举个例子，尝试清楚地理解这个概念。假设一个人住在城镇里，然后去购物商场买东西。一个产品有很多个。当他买东西时，我们可以生成他所买东西的模式。同样，我们可以生成不同人的销售和购买模式。现在你可能正在考虑一个人随机来买东西，然后他再也不会来了，但是我们也有此类事情的模式。在这种模式的帮助下，我们可以决定人们最喜欢的事物以及他们何时再次来到购物中心。他们将在入口处看到他们想要的东西。这就是我们通过机器学习吸引客户的方式。

3．机器如何学习？

实际上，机器通过数据模式学习。让我们从数据的数据集开始。我们提供给机器的输入称为 X，得到的响应为 Y。这里有三种学习类型。

1）监督学习
2）无监督学习
3）强化学习

3.1　监督学习

在监督学习中，我们了解不同情况（输入），并且知道这些情况的标签（输出）。在这里，我们已经了解了基本事实，因此我们仅关注函数（操作），因为它是这里最主要和最重要的事情（图 7-1）。

（图略）

此时，我们仅创建函数以获取输入的输出。并且，我们尝试创建处理数据的函数，并在大多数情况下尝试给出准确的输出（Y）（图 7-2）。

（图略）

因为我们从输入的已知值开始，所以我们可以验证模型并将其改进。

现在，我们以不同的数据集教我们的机器。此时，是在未知情况下进行检查并生成值的时候了。

注意：假设你为机器提供了某种数据集，现在可以根据该数据集训练模型。现在，根据你所提供的知识集从该模型获得结果。但是，让我们假设，如果你删除了该知识集中的现有项目或进行了更新，那就得不到结果，因为你修改了数据集。

3.2   无监督学习

它与监督学习完全不同。在这里,我们不了解不同情况的标签(输出)。此时,我们通过寻找相似性来训练带有模式的模型。然后这些模式成为聚类(图 7-3)。

聚类 = 收集相似模式的数据

然后,用该聚类分析和处理数据。

在无监督学习中,我们真的不知道输出是对还是错。因此,在这种情况下,系统会识别出模式并尝试计算结果,直到获得接近正确的值为止。

3.3   强化学习

这就像基于奖励的学习。基于奖励的例子是,假设你的父母会在你完成特定任务时给你奖励。因此,此时你知道必须完成此任务及所需的时间。开发人员可以在这里自行决定完成此任务将获得的奖励。

这也是面向反馈的学习。现在你正在执行一些任务,并在这些任务的基础上获得了反馈。如果反馈是正面的,则说明你做对了,可以自己改善工作。如果反馈是负面的,那么你也知道问题出在哪里以及如何更正。反馈来自其工作环境。

它使系统比无人监督的方案更优化,因为这里有一些线索,例如奖励或良好的反馈,可以使我们的系统更高效。

4. 机器学习的步骤

当我们开始教机器时,有一些机器学习的关键步骤。

4.1   收集数据

众所周知,机器学习是面向数据的。我们需要数据来指导我们的系统进行未来的预测。

4.2   准备输入数据

现在,你已经下载了数据,但是在输入数据时,需要确保数据的特定顺序,以使其被你的机器学习工具进行处理时有意义,例如 .csv 文件(逗号分隔值)。这是处理数据的最佳文件格式,因为逗号分隔的值在聚类中有很大帮助。

4.3   分析数据

现在,你正在查看数据中的模式,以便以更好的方式对其进行处理。你正在检查数据的异常值(范围和边界)。你还正在检查数据的新颖性(规格)。

4.4   训练模型

当你开发算法时,这是机器学习的主要部分,在该算法中,你将使用编码以构造完整的系统来处理输入并返回输出。

4.5　测试模型

在这里,你要检查从系统获取的值是否与所需结果相匹配。

4.6　在应用程序中部署

让我们讨论一个自动驾驶的示例,该示例无需人工干预,而是自行运行。第一步是收集数据,你必须收集多种数据。你驾驶的车自主运行。汽车应该知道路标,应该了解交通信号以及人们过马路的知识,以便可以做出在不同情况下停车或行驶的决定。因此,我们需要收集这些不同情况的图像,这就是我们的"收集数据模块"。

现在,我们必须制作特定格式的数据(图像),例如.csv 文件,在其中存储文件的路径及文件的大小。它使我们的系统处理效率更高。这就是我们所谓的准备数据。

然后,它为不同的交通信号(红色、绿色、黄色),不同的交通标志牌以及汽车或人员在附近奔波的环境制定模式。其次,确定这些对象的异常值是静态的(已停止)还是动态的(运行状态)。最后,它可以决定停止还是移到另一个对象的侧面。

这些决定显然取决于训练模型以及我们为开发模型编写的代码。这就是我们所说的训练模型及测试模型,然后将其部署到实际应用程序中的意思。

5. 机器学习的应用

如今,机器学习已在应用程序中广泛使用。

(1) 可以使用 Snapchat 或 Instagram 应用程序把动物的不同身体部分(例如耳朵、鼻子、舌头等)放到你的面部图像上。将这些不同的器官放在图像的正确位置,就是机器学习的一种应用。

(2) Google 广泛使用人工智能和机器学习。Google Lens 是一个应用程序。如果你通过 Google 镜头扫描任何东西,它就能告诉你此特定事物的属性和功能。

(3) Google Maps 也正在使用机器学习。例如,如果你正在看地图上的任何百货商店,有时它会告诉你物品的价格多少,有多贵。

Unit 8

录音

Text A

Artificial Neural Network

1. What Is an Artificial Neural Network?

An artificial neural network is a biologically inspired computational model that is patterned after the network of neurons present in the human brain. Artificial neural networks can also be thought of as learning algorithms that model the input-output relationship. Applications of artificial neural networks include pattern recognition and forecasting in fields such as medicine, business, pure sciences, data mining, telecommunications, and operations managements.

An artificial neural network transforms input data by applying a nonlinear function to a weighted sum of the inputs. The transformation is known as a neural layer and the function is referred to as a neural unit. The intermediate outputs of one layer, called features, are used as the input into the next layer. The neural network through repeated transformations learns multiple layers of nonlinear features (like edges and shapes), which it then combines in a final layer to create a prediction (of more complex objects). The neural net learns by varying the weights or parameters of a network so as to minimize the difference between the predictions of the neural network and the desired values. This phase where the artificial neural network learns from the data is called training (See Figure 8-1).

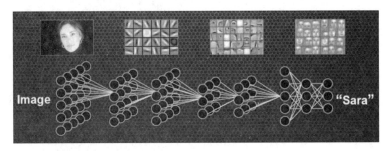

Figure 8-1   Schematic representation of a neural network

2. How Many Types of Neural Network Are There?

There are multiple types of neural network, each of which comes with their own specific use cases and levels of complexity. The most basic type of neural net is something called a feedforward neural network, in which information travels in only one direction from input to output.

A more widely used type of network is the recurrent neural network, in which data can flow in multiple directions. These neural networks possess greater learning abilities and are widely employed for more complex tasks such as learning handwriting or language recognition.

There are also convolutional neural networks, Boltzmann machine networks, Hopfield networks, and a variety of others. Picking the right network for your task depends on the data you have to train it with, and the specific application you have in mind. In some cases, it may be desirable to use multiple approaches, such as would be the case with a challenging task like voice recognition.

3. Neural Network Inference

Once the artificial neural network has been trained, it can accurately predict outputs when presented with inputs, a process referred to as neural network inference. To perform inference, the trained neural network can be deployed in platforms ranging from the cloud, to enterprise data centers, to resource-constrained edge devices. The deployment platform and type of application impose unique latency, throughput, and application size requirements on runtime. For example, a neural network performing lane detection in a car needs to have low latency and a small runtime application. On the other hand, data center identifying objects in video streams needs to process thousands of video streams simultaneously, needing high throughput and efficiency.

4. How Does a Neural Network 'Learns' Stuff Exactly?

In the same way that we learn from experience in our lives, a neural network requires data to learn. In most cases, the more data that can be thrown at a neural network, the more accurate it will become. Think of it like any task you do over and over. Over time, you gradually get more efficient and make fewer mistakes.

When researchers or computer scientists set out to train a neural network, they typically divide their data into three sets. First is a training set, which helps the network establish the various weights between its nodes. After this, they fine-tune it using a validation data set. Finally, they'll use a test set to see if it can successfully turn the input into the desired output.

5. What Kind of Tasks Can a Neural Network Do?

There are many kinds of tasks a neural network can do, from making cars drive autonomously on the roads to generating shockingly realistic CGI faces, to machine translation, to fraud detection, to reading our minds, to recognizing when a cat is in the garden and turning on the sprinklers; neural nets are behind many of the biggest advances in AI.

Broadly speaking, however, neural networks are designed for spotting patterns in data. Specific tasks could include classification (classifying data sets into predefined classes), clustering (classifying data into different undefined categories), and prediction (using past events to guess future ones, like the stock market or movie box office).

6. Accelerating Artificial Neural Networks with GPUs

State-of-the-art neural networks can have from millions to well over one billion parameters to adjust via back-propagation. They also require a large amount of training data to achieve high accuracy, meaning hundreds of thousands to millions of input samples will have to be run through both a forward and backward pass. Because neural nets are created from large numbers of identical neurons they are highly parallel by nature. This parallelism maps naturally to GPUs, which provide a significant computation speed-up over CPU-only training.

GPUs have become the platform of choice for training large，complex Neural Network-based systems because of their ability to accelerate the systems. Because of the increasing importance of Neural networks in both industry and academia and the key role of GPUs，NVIDIA has a library of primitives called cuDNN that makes it easy to obtain state-of-the-art performance with deep neural networks.

The parallel nature of inference operations also lend themselves well for execution on GPUs. To optimize，validate，and deploy networks for inference，NVIDIA has an inference platform accelerator called TensorRT. TensorRT delivers low-latency，high-throughput inference and tunes the runtime application to run optimally across different families of GPUs.

7. Do Neural Networks Have Any Limitations?

On a technical level，one of the bigger challenges is the amount of time it takes to train networks，which can require a considerable amount of compute power for more complex tasks. The biggest issue，however，is that neural networks are 'black boxes'，in which the user feeds in data and receives answers. They can fine-tune the answers，but they don't have access to the exact decision making process.

This is a problem a number of researchers are actively working on，but it will only become more pressing as artificial neural networks play a bigger and bigger role in our lives.

✎ New Words

neuron	['njʊərɒn]	n. 神经元，神经细胞
biologically	[ˌbaɪəʊ'lɒdʒɪkəlɪ]	adv. 生物学上地
inspire	[ɪn'spaɪə]	vt. 赋予灵感，启发，启迪；激励
pattern	['pætn]	n. 模式
		vt. 模仿
telecommunication	[ˌtelɪkəˌmjuːnɪ'keɪʃn]	n. 电信
nonlinear	['nɒn'lɪnɪəl]	adj. 非线性的
transformation	[ˌtrænsfə'meɪʃn]	n. 转换；变化
intermediate	[ˌɪntə'miːdɪət]	adj. 中间的，中级的
		n. 中间物，中间人
		vi. 调解；干涉
feedforward	['fiːdfɔːwəd]	n. 前馈
recurrent	[rɪ'kʌrənt]	adj. 递归的；周期性的，经常发生的；循环的
direction	[də'rekʃn]	n. 方向

possess	[pə'zes]	*vt.* 拥有；掌握，懂得
handwriting	['hændraɪtɪŋ]	*n.* 书法，手书；笔迹，字迹
convolutional	[ˌkɒnvə'luːʃnl]	*adj.* 卷积的
desirable	[dɪ'zaɪərəbl]	*adj.* 令人满意的；值得拥有的；可取的
impose	[ɪm'pəʊz]	*vi.* 利用；施加影响
latency	['leɪtənsɪ]	*n.* 延迟；潜伏
throughput	['θruːpʊt]	*n.* 吞吐量；流率
runtime	['rʌntaɪm]	*n.* 运行时间，运行期
simultaneously	[ˌsɪməl'teɪnɪəslɪ]	*adv.* 同时地
exactly	[ɪg'zæktlɪ]	*adv.* 精确地；确切地
shockingly	['ʃɒkɪŋlɪ]	*adj.* 令人震惊地，极度地
realistic	[ˌriːə'lɪstɪk]	*adj.* 逼真的；栩栩如生的
fraud	[frɔːd]	*n.* 欺诈；骗子；伪劣品；冒牌货
sprinkler	['sprɪŋklə]	*n.* 洒水器，自动喷水灭火装置
spot	[spɒt]	*v.* 认出，发现
predefine	['priːdɪ'faɪn]	*vt.* 预先确定；预定义
back-propagation	[bæk-prɒpə'geɪʃn]	*n.* 反向传播
identical	[aɪ'dentɪkl]	*adj.* 同一的，相同的
parallelism	['pærəlelɪzəm]	*n.* 平行；对应，类似
map	[mæp]	*vt.* 映射
accelerate	[ək'seləreɪt]	*v.* 加快，加速
primitive	['prɪmətɪv]	*adj.* 原始的
accelerator	[ək'seləreɪtə]	*n.* 加速器
low-latency	[ləʊ-'leɪtənsɪ]	*adj.* 低延迟的
high-throughput	[haɪ-'θruːpʊt]	*adj.* 高吞吐量的
tune	[tjuːn]	*n.* 曲调；调谐
		vt. 调整
considerable	[kən'sɪdərəbl]	*adj.* 相当大(或多)的；该注意的,应考虑的

✎ Phrases

artificial neural network	人工神经网络
computational model	计算模型
learning algorithm	学习算法
pattern recognition	模式识别
pure science	(区别于应用科学的) 纯科学
data mining	数据挖掘
nonlinear function	非线性函数

neural layer	神经层
neural unit	神经单元
nonlinear feature	非线性特征
neural net	神经网络
desired value	期望值
feedforward neural network	前馈神经网络
information travel	信息传播,信息传输
recurrent neural network	递归神经网络
language recognition	语言识别
convolutional neural network	卷积神经网络
Boltzmann machine network	玻尔兹曼机器网络
Hopfield network	霍普菲尔德网络
voice recognition	声音识别,语音识别
resource-constrained edge device	资源受限的边缘设备
lane detection	车道检测
be thrown at	被扔向
set out to	打算,着手
fraud detection	欺诈检测
stock market	股票市场;股票买卖;股票行情
movie box office	电影票房
forward pass	正推法
backward pass	逆推法,倒推法
deep neural network	深度神经网络,深层神经网络
black box	黑盒子,黑匣子
decision making process	决策过程

✍ Abbreviations

CGI (Computer-Generated Imagery)	计算机产生的图像

✍ Exercises

【Ex.1】 根据课文内容回答问题。

1. What is an artificial neural network?

2. What do applications of artificial neural networks include?

3. What is the most basic type of neural net?

4. What is a more widely used type of network?

5. What can the artificial neural network accurately do once it has been trained?

6. What platforms can the trained neural network be deployed in to perform inference?

7. When researchers or computer scientists set out to train a neural network, how many sets do they typically divide their data into? What are they?

8. What tasks can a neural network do?

9. Why have GPUs become the platform of choice for training large, complex Neural Network-based systems?

10. What is one of the bigger challenges for neural networks on a technical level?

【Ex.2】 把下列单词或词组中英互译。

1. latency   1. _____

2. artificial neural network   2. _____

3. computational model   3. _____

4. convolutional neural network   4. _____

5. data mining   5. _____

6. 语言识别   6. _____

7. 学习算法   7. _____

8. 非线性函数   8. _____

9. 模式识别   9. _____

10. n.前馈   10. _____

【Ex.3】 短文翻译。

Artificial Neural Network

An artificial neuron network (ANN) is a computational model based on the structure and functions of biological neural networks. Information that flows through the network affects the structure of the ANN because a neural network changes or learns based on that input and output.

ANNs are considered nonlinear statistical data modeling tools where the complex relationships between inputs and outputs are modeled or patterns are found.

An ANN has several advantages but one of the most recognized of these is the fact that it can actually learn from observing data sets. In this way, ANN is used as a random function approximation tool. ANN takes data samples rather than entire data sets to arrive at solutions, which saves both time and money. ANNs are considered fairly simple mathematical models to enhance existing data analysis technologies.

ANNs have three layers that are interconnected. The first layer consists of input neurons. Those neurons send data on to the second layer, which in turn sends the output neurons to the third layer.

Training an artificial neural network involves choosing from allowed models for which there are several associated algorithms.

【Ex.4】 将下列词填入适当的位置(每个词只用一次)。

gray	neural	pattern	environment	sets
brain	interact	fuzzy	combining	climate

Fuzzy Neural Networks

Fuzzy neural networks are software systems that attempt to approximate the way in which the human __1__ functions. They do this by utilizing two key research areas in computer science technology—__2__ logic software development and neural network processing architecture. Fuzzy logic software attempts to account for real-world __3__ areas in the decision making structure of computer software programs that go beyond simple yes or no choices. Artificial __4__ network design creates software nodes that imitate the functionality and complexity of how neurons __5__ in the human brain. Together, fuzzy logic and neural network design creates a neuro-fuzzy system that researchers use for experimentation on complex problems such as __6__ change, or to develop artificial intelligence robotics.

The key elements that make fuzzy neural networks unique from other types of computer processing are their ability at __7__ recognition given insufficient data to draw definitive conclusions, and the ability to adapt to the __8__ . Fuzzy neural networks utilize neural algorithms that are designed to change and grow as they encounter new data __9__ to process. They do this by approaching problems from two distinct points of view and __10__ the results into meaningful solutions to problems.

Text B

Difference Between Supervised and Unsupervised Machine Learning

Supervised learning and unsupervised learning are two core concepts of machine learning. Supervised learning is a machine learning task of learning a function that maps an input to an output based on the example input-output pairs. Unsupervised learning is the machine learning task of inferring a function to describe hidden structure from unlabelled data. The key difference between supervised and unsupervised machine learning is that supervised learning uses labeled data while unsupervised learning uses unlabeled data.

Machine learning is a field in Computer Science that gives the ability for a computer system to learn from data without being explicitly programmed. It allows to analyze the data and to predict patterns in it. There are many applications of machine learning. Some of them are face recognition, gesture recognition and speech recognition. There are various algorithms related to machine learning. Some of them are regression, classification and clustering. The most common programming languages for developing machine learning based applications are R and Python. Other languages such as Java, C++ and Matlab can also be used.

1. What is Supervised Learning?

In machine learning based systems, the model works according to an algorithm. In supervised learning, the model is supervised. First, it is required to train the model. With the gained knowledge, it can predict answers for the future instances. The model is trained using a labeled data set. When an outcome of sample data is given to the system, it can predict the result.

In supervised learning, there are algorithms for classification and regression. Classification is the process of classifying the labeled data. The model creates boundaries that separate the categories of data. When new data is provided to the model, it can categorize based on where the point exists. The K-nearest neighbors (KNN) is a classification model. Depending on the k value, the category is decided. For example, when k is 5, if a particular data point is near to eight data points in category A and six data points in category B, then the data point will be classified as A.

The regression is the process of predicting the trend of the previous data to predict the outcome of the new data. In regression, the output can consist of one or more continuous variables. Prediction is done using a line that covers most data points. The simplest regression model is a linear regression. It is fast and does not require tuning parameters such as in KNN. If the data shows a parabolic trend, then the linear regression model is not suitable.

Those are some examples of supervised learning algorithms. Generally, the results generated from supervised learning methods are more accurate and reliable because the input data is well known and labeled. Therefore, the machine has to analyze only the hidden patterns.

2. What Is Unsupervised Learning?

In unsupervised learning, the model is not supervised. The model works on its own to predict the outcomes. It uses machine learning algorithms to come to conclusions on unlabeled data. Generally, the unsupervised learning algorithms are harder than supervised learning algorithms because there is little information. Clustering is a type of unsupervised learning. It can be used to group the unknown data using algorithms. The k-mean and density-based clustering are two clustering algorithms.

K-mean algorithm places k centroid randomly for each cluster. Then each data point is assigned to the closest centroid. Euclidean distance is used to calculate the distance from the data point to the centroid. The data points are classified into groups. The positions for k centroids are calculated again. The new centroid position is determined by the mean of all points in the group. Again each data point is assigned to the closest centroid. This process repeats until the centroids no longer change. K-mean is a fast clustering algorithm，but there is no specified initialization of clustering points. Also，there is a high variation of clustering models based on initialization of cluster points.

Another clustering algorithm is density based clustering. It is also known as Density Based Spatial Clustering Applications with noise. It works by defining a cluster as the maximum set of density connected points. They are two parameters used for density based clustering. They are ε (epsilon) and minimum points. The ε is the maximum radius of the neighborhood. The minimum points are the minimum number of points in the ε neighborhood to define a cluster. Those are some examples of clustering that falls into unsupervised learning.

Generally，the results generated from unsupervised learning algorithms are not much accurate and reliable because the machine has to define and label the input data before determining the hidden patterns and functions.

3. What is the Difference Between Supervised and Unsupervised Machine Learning(See Table 8-1)?

Table 8-1   Difference Between Supervised and Unsupervised Machine Learning

Supervised vs Unsupervised Machine Learning	
Supervised learning is the machine learning task of learning a function that maps an input to an output based on example input-output pairs	Unsupervised learning is the machine learning task of inferring a function to describe hidden structure from unlabeled data
Main Functionality	
In supervised learning，the model predicts the outcome based on the labeled input data	In unsupervised learning，the model predicts the outcome without labeled data by identifying the patterns on its own

续表

Accuracy of the Results	
The results generated from supervised learning methods are more accurate and reliable	The results generated from unsupervised learning methods are not much accurate and reliable
Main Algorithms	
There are algorithms for regression and classification in supervised learning	There are algorithms for clustering in unsupervised learning

4. Summary

Supervised learning and unsupervised learning are two types of machine learning. Supervised learning is the machine learning task of learning a function that maps an input to an output based on example input-output pairs. Unsupervised learning is the machine learning task of inferring a function to describe hidden structure from unlabeled data. The difference between supervised and unsupervised machine learning is that supervised learning uses labeled data while unsupervised leaning uses unlabeled data.

✎New Words

supervise	['sju:pəvaɪz]	v. 监督；管理；指导
unsupervise	[ʌn'sju:pəvaɪz]	v. 无监督；无管理
pair	[peə]	n. 一对，一副
		v. (使……)成对，(使……)成双
unlabelled	[ʌn'leɪbld]	adj. 未标记的
gesture	['dʒestʃə]	n. 手势
		vt. 做手势
instance	['ɪnstəns]	n. 例子，实例；情况；要求，建议
sample	['sɑːmpl]	n. 样本，样品
		vt. 取……的样品；抽样调查
regression	[rɪ'greʃn]	n. 回归
continuous	[kən'tɪnjuəs]	adj. 连续的；延伸的；不断的
parabolic	[ˌpærə'bɒlɪk]	adj. 抛物线的
suitable	['sju:təbl]	adj. 合适的，适当的
analyze	['ænəlaɪz]	vt. 分析
clustering	['klʌstərɪŋ]	n. 聚类
centroid	['sentrɔɪd]	n. 质心；矩心
randomly	['rændəmlɪ]	adv. 随机地，随便地

euclidean	[juːˈklɪdɪən]	*adj*. 欧几里得的, 欧几里得几何学的
initialization	[ɪˌnɪʃəlaɪˈzeɪʃn]	*n*. 设定初值, 初始化
noise	[nɔɪz]	*n*. 噪音, 杂音

✍ Phrases

unlabelled data	未标记的数据
gesture recognition	手势识别
sample data	样本数据
regression model	回归模型
linear regression	线性回归
hidden pattern	隐藏模式, 隐含模式
density-based clustering	基于密度的聚类方法
fast clustering algorithm	快速聚类算法
Density Based Spatial Clustering Application	基于密度的空间聚类应用

✍ Abbreviations

| KNN (K-Nearest Neighbors) | K 最近邻算法 |

✍ Exercises

【Ex.5】 根据课文内容回答问题。

1. What is supervised learning?

2. What is unsupervised learning?

3. What are some of the various algorithms related to machine learning?

4. What are the most common programming languages for developing machine learning based applications?

5. What is regression?

6. Why are the results generated from supervised learning methods are more accurate and reliable generally?

7. Why are the unsupervised learning algorithms harder than supervised learning algorithms generally?

8. How does density based clustering work?

9. Why are the results generated from unsupervised learning algorithms not much accurate and reliable generally?

10. How does the model predict the outcome in unsupervised learning?

Reading

Robots

1. Different Branches Occupied in the Development of Robotics

Robotics, in contrast to other branches, is a reasonably new domain of engineering. Itisa multi-disciplinary① domain. The different branches occupied in the development of Robotics are:

(1) Mechanical Engineering: Deals with the machinery and structure of the robots.

(2) Electrical Engineering: Deals with the controlling and intelligence(sensing) of robots.

(3) Computer Engineering: Deals with the movement, development and observation of robots.

2. Classification of Robots

Robots are categorized depending upon the circuits of the robots and the variety of application it can perform. The robots are classified into three types:

(1) Simple level robots—These are automatic machines which do not contain complex circuit. They are developed just to extend human potential, for example, washing machine.

(2) Middle level robots—These robots are programmed but can never be reprogrammed②. These robots contain sensor based circuit & can perform multiple tasks, for example, fully automatic washing machine.

(3) Complex level robots—These robots are programmed and can be reprogrammed as well. They contain complex model based circuit, for example, laptop or computer.

3. Types of Robotics

Robotics is an area of interest to human beings for more than one hundred years.

---

① multi-disciplinary [ˌmʌltɪˈdɪsəplɪnerɪ] adj. 多学科的

② reprogrammed [ˈriːprəugræmd] v. 再次编程，程序重调

On the other hand, our perception over robots is influenced by the media and international film industry (Hollywood). You may ask what robotics is all about. In my views, a robot's distinctiveness① transforms depending on the atmosphere② it works in. Some of these are as follows:

(1) Outer space—Robotic arms that are under the control of a human being are employed to unload the docking cove of outer-space shuttles③ to launch satellites or to build a space station④.

(2) The intelligent home—Robotic systems can nowadays scrutinize⑤ home safety, ecological circumstances⑥ and energy consumption⑦. Doors and windows can be unlocked mechanically and electrical device such as lights and A/C can be pre-programmed to turn on. This helps residents to enjoy appliances irrespective⑧ of their mobility.

(3) Exploration—Robots can enter the environments that are injurious⑨ to human beings. An illustration is observing the atmosphere within a volcano⑩ or investigating our deep marine life. NASA has utilized robotic probe for environmental study, ever since the early 1960's.

(4) Military Robots—Flying robot drones are brought into play for close watch in present time's modern armed force. In the future robotic airplane and automobiles could be employed to transmit petroleum, bullets, bombs, etc or clear minefields⑪.

(5) Farms—Programmed robots are used by harvesters⑫ to cut and collect crops. Robotic milk farms are existing permitting workers to nourish and milk their cattle distantly.

(6) The car industry—Robotic arms are used. These arms are able to execute numerous tasks in the car manufacturing and assembling⑬ procedure. They carry out jobs such as sorting, cutting, welding⑭, lifting, painting and bending. Similar functions

---

① distinctiveness [dɪ'stɪŋktɪvnɪs] n.独特性
② atmosphere ['ætməsfɪə] n.风格,基调
③ shuttle ['ʃʌtl] n.航天飞机
④ space station:空间站
⑤ scrutinize ['skru:tənaɪz] vt.仔细检查
⑥ ecological circumstances:生态环境
⑦ consumption [kən'sʌmpʃn] n.消耗,消费
⑧ irrespective [ˌɪrɪ'spektɪv] adj.不考虑的,不顾的;无关的
⑨ injurious [ɪn'dʒʊərɪəs] adj.有害的
⑩ volcano [vɒl'keɪnəʊ] n.火山
⑪ minefield ['maɪnfi:ld] n.布雷区
⑫ harvester ['hɑ:vɪstə] n.收割机
⑬ assembling [ə'semblɪŋ] v.装配,组合,组装
⑭ welding [weldɪŋ] v.焊接

but on a minor scale are now being intended for the food industry to execute tasks like the trimming, cutting and processing of different types of meats like chicken, beef, fish, lamb, etc.

(7) Hospitals—The development of a robotic suit is under construction that will allow nurses to raise patients without injuring their backbones[①]. Scientists in Japan have crafted a power facilitated suit which will provide nurses the additional power that they need to lift patients.

(8) Disaster areas[②]—Observation robots built-in with superior sensing and imaging gears. This robot can work in dangerous environments like urban site spoiled by earthquakes by inspecting floors, walls, and roofs for structural reality.

(9) Entertainment[③]—Interactive robots[④] that shows behaviors and education capability. One such robot is owned by SONY which moves around freely, responds to all your commands, carries your luggage and even responds to your oral instructions[⑤].

This is not the end of robotic world. There are many more applications of Robotics.

4. Applications

Currently, robots perform a number of different jobs in numerous fields and the amount of tasks delegated to robots is rising progressively[⑥]. The best way to split robots into types is a partition by their application.

(1) Industrial robots—These robots bring into play in an industrialized manufacturing atmosphere. Typically these are articulated arms particularly created for applications like material handling, painting, welding and others. If we evaluate merely by application then this sort of robots can also consist of some automatically guided automobiles and other robots.

(2) Domestic[⑦] or household robots—Robots which are used at home. This sort of robots consists of numerous different gears for example, robotic pool cleaners, robotic sweepers, robotic vacuum[⑧] cleaners, robotic sewer cleaners and other robots that can

---

① backbone ['bækbəʊn] n.脊梁骨,脊椎
② disaster area：灾区
③ entertainment：[,entə'teɪnmənt] n.娱乐,消遣
④ interactive robot：交互式机器人
⑤ oral instruction：口头指令
⑥ progressively [prə'gresɪvlɪ] adv.日益增加地,逐步地
⑦ domestic [də'mestɪk] adj.家庭的
⑧ vacuum ['vækjuəm] n.真空 v.用真空吸尘器清扫

perform different household tasks. Also, a number of scrutiny and tele-presence robots[①] can also be considered as domestic robots if brought into play in that sort of environment.

(3) Medical robots—Robots employed in medicine and medicinal institutes. First and foremost surgical treatment robots. Also, a number of robotic directed automobiles and perhaps lifting supporters.

(4) Service robots—Robots that cannot be classed into any other types by practice. These could be various data collecting robots, robots prepared to exhibit[②] technologies, robots employed for research, etc.

(5) Military robots—Robots brought into play in military & armed forces. This sort of robots consist of bomb discarding robots[③], various shipping robots, exploration drones. Often robots at the start produced for military and armed forces purposes can be employed in law enforcement[④], exploration and salvage and other associated fields.

(6) Entertainment robots—These types of robots are employed for entertainment. This is an extremely wide ranging category. It begins with model robots such as Robosapien or the running photo frames and concludes with real heavy weights like articulated robot arms employed as movement simulators.

(7) Space robots—I would like to distinct out robots employed in space as a split apart type. This type of robots would consist of the robots employed on Canadarm that was brought into play in space shuttles, the International Space Station, together with Mars explorers and other robots employed in space exploration and other activities.

(8) Hobby and competition robots[⑤]—Robots that is created by students. Sumo-bots, Line followers, robots prepared merely for learning, fun and robots prepared for contests.

Now, as you can see that there are a number of examples that fit well into one or more of these types. For illustration, there can be a deep ocean discovery robot that can collect a number of precious information that can be employed for military or armed forces purpose.

Robotics is a broad field and everyday there is a pioneering[⑥] invention in the field. Robots were invented by the humans just for fun but by now they are used for

---

① tele-presence robot：远距临场机器人

② exhibit [ɪɡˈzɪbɪt] vt．陈列，展览；呈现

③ bomb discarding robot：拆弹机器人

④ enforcement [ɪnˈfɔːsmənt] n．强制，实施，执行

⑤ competition robot：竞赛机器人

⑥ pioneering [ˌpaɪəˈnɪərɪŋ] adj．首创的，先驱的，开创性的

assisting humans in various sectors. Human beings are better suitable for multifaceted①, imaginative, adaptive jobs, and robots are good for dreary②, recurring tasks, permitting human beings to do the harder thinking jobs, whereas a robot is employed for substituting humans for various recurring tasks or entertainment to make living more expedient.

参考译文

人工神经网络

1. 什么是人工神经网络？

人工神经网络是一种受生物学启发的计算模型，它是人脑中存在的神经元网络的模型化。人工神经网络也可以被认为是模拟输入—输出关系的学习算法。人工神经网络应用于医学、商业、纯科学、数据挖掘、通信和运营管理等领域的模式识别和预测。

人工神经网络通过将非线性函数应用于输入的加权和来转换输入数据。该变换称为神经层，该函数称为神经元。一层的中间输出（称为特征）用作下一层的输入。通过重复变换的神经网络学习多层非线性特征（如边缘和形状），然后在最后一层中组合以创建预测（更复杂的对象）。神经网络通过改变网络的权重或参数来学习，以使神经网络的预测与期望值之差最小化。人工神经网络从数据中学习的这个阶段称为训练（参看图 8-1）。

2. 有多少种神经网络？

有多种类型的神经网络，每种神经网络都有自己的特定用例和复杂程度。最基本类型的神经网络称为前馈神经网络，其中信息仅在一个方向上从输入传到输出。

更广泛使用的网络类型是递归神经网络，其中数据可以在多个方向上流动。这些神经网络具有更强的学习能力，并广泛用于更复杂的任务，如学习识别手写文字或识别语言。

还有卷积神经网络、玻尔兹曼机器网络、霍普菲尔德网络以及其他各种网络。根据你必须使用的数据以及考虑的具体应用程序来为你的任务选择合适的网络。在某些情况下（例如像语音识别这样具有挑战性的任务）可能需要使用多种方法。

---

① multifaceted [ˌmʌltɪˈfæsɪtɪd] *adj*. 多方面的，多才多艺的
② dreary [ˈdrɪərɪ] *adj*. 令人厌烦的，枯燥的

3．神经网络推理

一旦人工神经网络经过训练，它就可以在输入时准确地预测输出，这一过程称为神经网络推理。为了进行推理，训练好的神经网络可以部署在从云、企业数据中心到资源有限的边缘设备的平台中。部署平台和应用程序类型对运行时间有特殊的延迟、吞吐量和应用程序大小的要求。例如，在汽车中执行车道检测的神经网络需要延迟低和运行快。另一方面，识别视频流中的对象的数据中心需要同时处理数千个视频流，这需要高吞吐量和高效率。

4．神经网络如何正确地"学习"东西？

就像我们从生活中的经验中学习一样，神经网络需要数据来学习。在大多数情况下，给神经网络的数据越多，它就越准确。把它想成你一遍又一遍执行的任务。随着时间的推移，你的效率越来越高，错误越来越少。

当研究人员或计算机科学家开始训练神经网络时，他们通常将数据分成三组。首先是训练集，它有助于网络在其节点之间建立各种权重。在此之后，他们使用验证数据集对其进行微调。最后，他们将使用测试集来查看它是否能够成功地将输入转换为所需的输出。

5．神经网络可以执行哪些任务？

神经网络可以做很多种任务，包括自动驾驶道路上的汽车，生成极为逼真的 CGI 面孔，机器翻译，欺诈检测，了解我们的想法，知道猫在花园里的时间和打开洒水器。神经网络是人工智能重大的支撑。

然而，从广义上讲，神经网络的设计用于识别数据中的模式。具体任务可以包括分类（将数据集分类为预定义的类）、聚类（将数据分类为不同的未定义类别）和预测（使用过去的事件来猜测未来的事件，如股票市场或电影票房）。

6．利用 GPU 加速人工神经网络

最先进的神经网络可以通过反向传播参数进行调整，这些参数的量从数百万到超过 10 亿。它们还需要大量的训练数据才能实现高精度，这意味着必须通过前向和后向传递数十万到数百万的输入样本。因为神经网络是由大量相同的神经元产生的，所以它们本质上是高度并行的。这种并行性自然地映射到 GPU，GPU 的计算速度比仅仅训练 CPU 高得多。

GPU 已成为训练大型复杂神经网络系统的首选平台，因为它们能够加速系统。由于神经网络在工业界和学术界的重要性日益增加以及 GPU 的关键作用，NVIDIA 拥有一

个名为 cuDNN 的原始库,可以很容易地利用深度神经网络获得最先进的性能。

推理操作的并行性质也很适合在 GPU 上执行。为了优化、验证和部署网络以进行推理,NVIDIA 有一个名为 TensorRT 的推理平台加速器。TensorRT 提供低延迟、高吞吐量推理,并调整运行期应用程序,以便在不同的 GPU 系列中实现最佳运行。

7. 神经网络有任何限制吗?

在技术层面上,一个更大的挑战是培训网络所需的时间,这可能需要相当大的计算能力来完成更复杂的任务。然而,最大的问题是神经网络是"黑匣子",用户在其中输入数据并接收答案。他们可以微调答案,但他们无法访问确切的决策过程。

这是许多研究人员正在积极研究的问题,但随着人工神经网络在我们的生活中发挥越来越大的作用,它将变得更加紧迫。

Unit 9

录音

Text A

Artificial Intelligence Trends to Be on High Demand in the Near Future

The year 2018 witnessed a dramatic rise in artificial intelligence platforms and applications. Along with the software and internet industry, the technology took other verticals like healthcare, legal, manufacturing, automobile and agriculture by storm. Earlier, artificial intelligence was restricted to science fiction movies, but today, we can see technology has caught up with imagination. A dream comes true. AI has become a reality. Every single person we encounter is accustomed to using AI in his or her everyday lives. A team of experts reported the rise of AI to be an industrial revolution at par with the three industrial revolutions of steam, oil and electricity, and computers.

Soon we will find AI revolution to be the gateway transforming cities into 'informational infrastructure' nestled in the broader 'digital revolution'. A recent report says 23% of businesses have incorporated the technology into its processes. Market research says the global artificial intelligence software market is expected to rise to 118.6 billion by 2025.

According to a recent survey, the significant improvements and game-changers in

AI are just a few months away. The AI trends in the near future will undoubtedly be the driving forces in both business and society.

Let us see the AI trends that will make a dent in the near future.

1. Chatbots

The skill to process natural language is broadly considered a hallmark of intelligence. The advancements in artificial intelligence have led to chatbot replacing traditional conversational service. Since chatbots are trained based on past information, organizations use them to maintain reports. The technology utilizes these logs to figure out client queries. With a blend of history and machine learning tools, the clients will get the appropriate answer.

The chatbots are based on three classification models.

1.1 Pattern Matching

The technology uses artificial intelligence markup language to group the text and produce an appropriate response. The chatbots try to relate the query with pattern combinations present in the database.

1.2 Natural Language Understanding (NLU)

NLU algorithm analyses the human query independent of past conversations. To differentiate the earlier conversations, its state is stored. It may be restaurants, office, etc. These parameters are much more helpful to relate the query with the context without having to worry about the chat history.

1.3 Natural Language Processing (NLP)

NLP converts the text to structured data. It uses a combination of the below-given steps:

(1) Tokenization: The sentences or series of words are separated into tokens or pieces that are linguistically representative.

(2) Sentiment Analysis: NLP studies the human experience to provide a satisfying reply to the human.

(3) Normalization: The text is processed to detect the typographical errors and spelling mistakes that alter the meaning of user query.

(4) Named Entity Recognition: The program model searches for specific words like name, address, etc. relating to the query.

(5) Dependency Parsing: It looks for subjects, verbs, common phrases, nouns and objects in the user's text to find out the phrases related to the user's request.

The AI-driven chatbots can handle complex human interactions with ease. Businesses have adopted these AI-enabled chatbots to deliver a personalized experience to the clients. The chatbots are designed to handle human queries and guide the clients through the complex procedures with ease.

2. Virtual Assistants(VA)

The digital landscape is currently experiencing a trend where virtual assistance is being acknowledged since the last decade. The recent years witnessed a phenomenal growth in the acceptance of virtual assistants like Siri, Alexa, Google assistant, and so on. The global market scenario states the growing need for multitasking ability for organizational success has led to the acceptance of VA. It is estimated that the number of people using virtual assistants across the globe is projected to reach 1.8 billion by 2021. These virtual assistants present on most of the smartphones understand the voice commands from the user and execute the task as requested. They are a combination of chips, microphones, and software that recognize the voice query from you and respond with the voice you select.

Virtual assistants have found a new home in a business environment. A survey says organizations are using virtual assistants to improve efficiency. Here are the uses of VA in businesses.

2.1 Automated meeting coordination

If a virtual assistant can access the organizational calendar of all the employees and clients, then it can book the meeting slots depending on everyone's schedule. A few virtual assistants notify external attendees in case of the meeting being cancelled.

2.2 Connecting with customers

Virtual assistants enable organizations to interact directly with the customers. The localized actions help the companies to customize the content and experience depending on location, language and region. Also, they help to develop compatible apps that can gather customer data to improve sales and outreach.

2.3 Managing infrastructure

The voice assistants are capable of monitoring backend computing services of an organization. They can monitor the resources available, active security alarms and the tasks performed.

2.4　Enabling the smart office

Voice assistants have made the concept of intelligent offices much more manageable. The digital assistants can control the amenities like lights, projectors, etc. in the room, depending on its usability. They also generate alerts in case of unavailability of resources.

Virtual assistants have made human life much more comfortable. We've started relying on voice assistants to play radio, to receive weather forecast, read emails, phone calls, to receive traffic news, etc.

3. Facial Recognition

The advances in digital sensors, processing power, data analytics and neural networks have given rise to stunning accuracy in the FR system. Face recognition system uses an LBPH algorithm to detect the human face correlating the contours of the eyes, nose, lips, ears and chin. Face recognition algorithm can be set in a way that they can even gauge emotions.

Face recognition system operates in two modes:

3.1　Verification or authentication of a facial image

In this mode, there is a comparison between input facial image and the image required for authentication. It is generally one to one correlation.

3.2　Identification or facial recognition

The input facial image is compared with all the images in the data set to find the match. The comparison is 1xN.

4. AI-enabled chips

The increasing acceptance of deep learning and machine learning models have demanded the need to use powerful chips for crunching enormous numbers. The technological advancements in the past few years have pushed the AI chips to emerge victorious than before. The following years will see the rise of specialized chips, manufactured to perform complicated mathematical computations and speed up the execution of AI-enabled applications. These AI-enabled chips will speed up tasks like object detection and facial recognition. AI-enabled chips on smartphones offer high data privacy and security.

These chips are optimized for specific use cases and scenarios regarding computer vision，natural language processing and speech recognition. The industries in the next generation will depend on these chips to deliver intelligence to end-users.

5. Conclusion

In the near future，the AI trends are sure to make a big splash in the market. The advancements in AI are not going to decline anytime soon. AI is not only accepted in the software and internet industry but has gained popularity in healthcare，automobile，retail sector etc. It will ease business operations and human lives. A brighter future awaits everyone，where humans are relieved of menial work using AI capabilities.

✎ New Words

witness	['wɪtnɪs]	vt.表示，提供……的证据
dramatic	[drə'mætɪk]	adj.戏剧性的；引人注目的；突然的；巨大的
platform	['plætfɔːm]	n.平台
software	['sɒftweə]	n.软件
automobile	['ɔːtəməbiːl]	n.汽车
imagination	[ɪˌmædʒɪ'neɪʃn]	n.想象，想象力
gateway	['geɪtweɪ]	n.门；入口；途径
process	['prəʊses]	n.过程
		vt.加工；处理
survey	['sɜːveɪ]	vt.调查；勘测
		n.调查(表)
game-changer	[geɪm-'tʃeɪndʒə]	n.规则改变者；打破格局、扭转局面的事物
undoubtedly	[ʌn'daʊtɪdlɪ]	adv.毋庸置疑地，的确地；显然地；必定地
hallmark	['hɔːlmɑːk]	n.特点，标志
		vt.使具有……标志
advancement	[əd'vɑːnsmənt]	n.前进，进步；提升，升级
log	[lɒg]	n.记录；日志
matching	['mætʃɪŋ]	adj.相配的；一致的；相称的
		n.匹配
analyse	['ænəlaɪz]	vt.分析；分解
conversation	[ˌkɒnvə'seɪʃn]	n.交谈，会话；(人与计算机的)人机对话
context	['kɒntekst]	n.上下文；背景；语境
worry	['wʌrɪ]	n.烦恼，忧虑；担心
		v.担心，焦虑，发愁

convert	[kən'vɜːt]	v.转变
text	[tekst]	n.文本
tokenization	[ˌtəʊkəaɪ'zeɪʃn]	n.词语切分
token	['təʊkən]	n.记号
		adj.作为标志的
representative	[ˌreprɪ'zentətɪv]	n.代表
		adj.典型的；有代表性的
satisfy	['sætɪsfaɪ]	v.使满意,满足
normalization	[ˌnɔːməlaɪ'zeɪʃn]	n.规范化,正常化,标准化
parse	[pɑːz]	vt.从语法上描述或分析(词句等)
adopt	[ə'dɒpt]	vt.采用,采取
client	['klaɪənt]	n.顾客；当事人；[计算机]客户端
procedure	[prə'siːdʒə]	n.程序；过程,步骤
assistant	[ə'sɪstənt]	n.助手,助理
		adj.助理的；辅助的
phenomenal	[fə'nɒmɪnl]	adj.显著的
acceptance	[ək'septəns]	n.接受,接纳
multitask	[ˌmʌltɪ'tɑːsk]	n.多任务
smartphone	[smɑːtfəʊn]	n.智能手机
chip	[tʃɪp]	n.芯片
microphone	['maɪkrəfəʊn]	n.麦克风,话筒
coordination	[kəʊˌɔːdɪ'neɪʃn]	n.协调
calendar	['kælɪndə]	n.日程表
attendee	[ˌæten'diː]	n.(会议等的)出席者
cancel	['kænsl]	vt.取消,注销
outreach	['aʊtriːtʃ]	n.扩大服务范围
		adj.扩大服务的
alarm	[ə'lɑːm]	n.警报
		vt.警告
manageable	['mænɪdʒəbl]	adj.可管理的,易处理的,易控制的
amenity	[ə'miːnətɪ]	n.便利设施
usability	[ˌjuːzə'bɪlɪtɪ]	n.可用性；适用性
unavailability	['ʌnəˌveɪlə'bɪlɪtɪ]	n.无效用,不适用
comfortable	['kʌmftəbl]	adj.舒适的
email	['iːmeɪl]	n.电子邮件
		vt.给……发电子邮件
sensor	['sensə]	n.传感器
gauge	[geɪdʒ]	n.评估

		vt. 评估，判断
verification	[ˌverɪfɪˈkeɪʃn]	n. 核实；证实
authentication	[ɔːˌθentɪˈkeɪʃn]	n. 身份验证；认证
correlation	[ˌkɒrəˈleɪʃn]	n. 相互关系；相关性
crunch	[krʌntʃ]	vt. 快速处理
victorious	[vɪkˈtɔːrɪəs]	adj. 胜利的，得胜的
execution	[ˌeksɪˈkjuːʃn]	n. 执行，完成
splash	[splæʃ]	v. (使)溅起；引人注目
		n. 扑通声
skilful	[ˈskɪlfl]	adj. 灵巧的；熟练的；技术好的

✎ Phrases

trend to	趋向，趋于
along with	和……一起，随着；以及；连同
internet industry	互联网产业
be restricted to	仅限于
science fiction movie	科幻电影
catch up with	追上，赶上
be accustomed to	习惯于……
informational infrastructure	信息基础设施
digital revolution	数字革命
a blend of	混合
markup language	标识语言
structured data	结构化数据
sentiment analysis	情感分析
typographical error	印刷错误，排字错误
spelling mistake	拼写错误
common phrase	常见短语，常用短语
personalized experience	个性化体验
voice command	语音命令
be capable of	能够
smart office	智能办公室
voice assistant	语音助理
weather forecast	天气预报
gauge emotion	判定情绪
be compared with...	与……相比较
AI-enabled chip	人工智能芯片

speed up	加速；增速
data privacy	数据保密
retail sector	零售部门

✎ Abbreviations

NLU (Natural Language Understanding)	自然语言理解
NLP (Natural Language Processing)	自然语言处理
VA (Virtual Assistant)	虚拟助手
FR (Facial Recognition)	人脸识别，面部识别
LBPH (Local Binary Pattern Histogram)	局部二值模式直方图

✎ Exercises

【Ex. 1】 根据课文内容填空。

1. The year 2018 witnessed a dramatic rise in _____ and _____.

2. Market research says the global artificial intelligence software market is expected to rise to _____ by 2025.

3. Since chatbots are trained based on past information，organizations use them to _____. The chatbots are based on three classification models：_____，_____ and _____.

4. NLU algorithm analyses the human query _____. To differentiate the earlier conversations，_____ is stored.

5. Tokenization：The sentences or _____ are separated into tokens or pieces that are _____.

6. It is estimated that the number of people using virtual assistants across the globe is projected to reach 1.8 billion by _____. These virtual assistants present on most of the smartphones _____ from the user and _____ as requested.

7. If a virtual assistant can access _____ of all the employees and clients，then it can book the meeting slots _____.

8. The digital assistants can control the amenities like _____，_____, etc. in the room，depending on its usability. They also _____ in case of unavailability of resources.

9. Face recognition system uses an LBPH algorithm to detect _____ correlating the contours of the eyes，nose，lips，ears and chin. Face recognition algorithm can be set in a way that they can even _____. Face recognition system operates in two modes：_____ and _____.

10. These AI-enabled chips will speed up tasks like _____ and _____.

AI-enabled chips on smartphones offer _____ and _____. These chips are optimized for specific use cases and scenarios regarding computer vision, _____ and speech recognition.

【Ex.2】 把下列单词或词组中英互译。

1. AI-enabled chip   1. _____
2. data privacy   2. _____
3. markup language   3. _____
4. structured data   4. _____
5. voice assistant   5. _____
6. n.身份验证；认证   6. _____
7. n.芯片   7. _____
8. n.交谈,会话   8. _____
9. n.相互关系；相关性   9. _____
10. n.多任务   10. _____

【Ex.3】 短文翻译。

Pervasive Computing

Pervasive computing is based on that technology is moving beyond the personal computer to everyday devices with embedded technology and connectivity as computing devices become progressively smaller and more powerful. Also called ubiquitous computing, pervasive computing is the result of computer technology advancing at exponential speeds—a trend toward all man-made and some natural products having hardware and software. Pervasive computing goes beyond the realm of personal computers: it is the idea that almost any device, from clothing to tools to appliances to cars to homes to the human body to your coffee mug, can be imbedded with chips to connect the device to an infinite network of other devices. The goal of pervasive computing, which combines current network technologies with wireless computing, voice recognition, Internet capability and artificial intelligence, is to create an environment where the connectivity of devices is embedded in such a way that the connectivity is unobtrusive and always available.

【Ex.4】 将下列词填入适当的位置(每个词只用一次)。

possess	intuitive	questions	nature	considered
mathematician	named	terminal	respondents	decide

Turing Test

A Turing Test is a method of inquiry in artificial intelligence (AI) for determining whether or not a computer is capable of thinking like a human being. The test is __1__ after Alan Turing, the founder of the Turning Test and an English computer scientist, cryptanalyst, __2__ and theoretical biologist.

Turing proposed that a computer could be said to __3__ artificial intelligence if it could mimic human responses under specific conditions. The original Turing Test requires three terminals, each of which is physically separated from the other two. One __4__ is operated by a computer, while the other two are operated by humans.

During the test, one of the humans functions as the questioner, while the second human and the computer function as respondents. The questioner interrogates the __5__ within a specific subject area, using a specified format and context. After a preset length of time or number of questions, the questioner is then asked to __6__ which respondent is human and which is a computer.

The test is repeated many times. If the questioner makes the correct determination in half of the test runs or less, the computer is __7__ to have artificial intelligence because the questioner regards it as 'just as human' as the human respondent.

The Turing Test has been criticized over the years, in particular because historically, the __8__ of the questioning had to be limited in order for a computer to exhibit human-like intelligence. For many years, a computer might only score high if the questioner formulated the queries, so they had 'Yes' or 'No' answers or pertained to a narrow field of knowledge. When __9__ were open-ended and required conversational answers, it was less likely that the computer program could successfully fool the questioner.

To many researchers, the question of whether or not a computer can pass a Turing Test has become irrelevant. Instead of focusing on how to convince someone they are conversing with a human and not a computer program, the real focus should be on how to make a human-machine interaction more __10__ and efficient. For example, by using a conversational interface.

Text B

How Can Artificial Intelligence Impact Cloud Computing

Cloud computing has always been a great technology that is transforming the way systems work, information stored and altering decisions paving their way for innovation and research. Imagine cloud getting smart? A smarter cloud can not only store and retrieve huge heaps of data, it can gather, disseminate and learn from that information as well to come with informed and instant decisions.

1. Intelligent Cloud

Working at the very basic level, AI has changed the way of data input, storage and analysis. Now, the cloud is not just a data storehouse but, it is an intelligent storehouse. Machine learning and the cloud computing together can save, analyze and learn at the same time from information and pass it on to other servers or clouds to help in framing information and response backed decision making. We can also hope for an intelligent cloud that's capable of predicting trends based on the user data inputs.

2. The Story So Far

The immense potential behind a cloud and artificial intelligence is known to everyone.

With the emerging business sphere embracing cloud computing and artificial intelligence in its core working style, there is a lot of scope in the future for the trend to keep motivating people and researchers to come up with better and intelligent cloud computing facilities. Technicians are working towards building platforms which can self-assess and self-decide minute details and come up with solutions for difficulties during the operational activities.

However, the AI backed cloud computing is still in its nascent stage. Companies would have to keep an eye on the developments in this arena to best utilize this technological miracle for their best interest. In recent years, we have seen mobiles replacing computers, IoT connecting everything together and AI empowering the intelligence. These changes have showered great impact on the cloud computing and technology, but to watch how AI will impact cloud computing would be really thrilling for all of us.

"The fusion of AI and cloud computing promises to be both a source of innovation and a means to accelerate change", is believed by IBM, one of the most appreciated cloud companies.

3. Hand in Hand Working

The cloud is really a fantastic source of information that keeps the learning mechanism of AI sound, while the AI can help provide useful responses and data analytics that can make the cloud computing result and situation oriented. Considering the vast scope of development in a symbiotic relation between AI and cloud computing, the rigorous efforts of IBM like organizations show that this is the big picture of future where AI and cloud computing will keep on creating amazing results.

4. Hopes for Futuristic Technology

Artificial intelligence is helping build machines or co-bots that can respond and understand the human behavior within a fraction of seconds. The machine learning capabilities of these bots are based on the wise analytical research done during the training or learning phase itself. This becomes well versed over the time to interact and come up with a quick response just like humans. For example, a bot named Tobi was used by Indian telco giant Vodafone to address consumers and provide them the repetitive kind of information about company's products and services. Windows 'Cortana' application is also a self-learning mechanism which is trained to observe human interaction and provide a useful response in real time.

5. Role of Human Beings in Shaping the Right AI and Cloud Collaboration

Due to the constantly changing scenarios and updation of technological upliftment, the learning need for AI has increased in demand. The cloud offers better data to let the machine learning accelerate without any type of interference. The human mind can redirect AI towards reading the information the right way. The cloud can also be well versed in providing useful data if AI is kept on radar via human technical experts.

6. Business Intelligence, Cloud and Smarter Workspaces

Businesses can make great leaps in decision making through the artificial intelligence by smart data storage and usage with the help of cloud computing. Companies can study past data to formulate business strategies, create future plans and simply analyze the information to understand the shortcomings during a weak critical period. Business forecasts done through AI can help millions of businesses derive strength and insights to carry on by putting in tremendous efforts at the right time.

7. Increased Demand for Cloud

In future, a cloud is going to become one amongst the top priorities that would protect from the challenges those businesses face. The need for huge data for the intelligence input would be provided by cloud or the 'Intelligent Cloud'. With cutting edge competition in the industrial sphere, having an intelligent cloud will be a necessity, not a choice anymore. Combining two big forces like AI and cloud computing will accelerate businesses for betterment. It would be more than appropriate to say that fields like healthcare, education, business, retail etc. will see an increasing demand for the AI infused intelligent clouds.

8. Increasing Application of Intelligent Cloud in Various Sectors

In the educational sector, an intelligent cloud comes into picture to provide data backed research studies and can guide students to take advantage of previous research and its implications. Likewise, health sector can do a lot with the intelligence of cloud. Doctors operating on a patient can seek reviews from thousands of similar cases done before. This by far beats the manual information being preserved and assessed by any medical practitioner. Elaborate data gathering, comparisons and timely presentation of new ways to solve complex medical surgeries, makes cloud a gift for medical industry. Areas like banking, investments and education can see more future friendly innovations embracing intelligence of AI and cloud's capabilities.

9. Data Learning Mechanism

There would be an increasing paradigm shift from basic level machine learning to the deep learning. In this scenario, AI-based cloud apps and computing algorithms make use of past data and research analytics to formulate future strategies and response. In other ways, you can expect bots and machines to interact and respond just like a human in real scenarios and situations harnessing the power of smart data and intelligent cloud computing.

10. Co-bots, Advanced Robotics, Personal Assistants

We have seen companies like Google and Microsoft come up with chatbots or personal assistants that can use the previous data inputs and can derive general knowledge from there. This makes human interaction more interesting and life easier for the people looking for repetitive information. The Google Alexa and Microsoft Cortana have artificial intelligence to come up with cloud computing based information. However, these devices are more generic in nature but will see tremendous operational and business scope with the advent of advanced cloud computing operations backed with artificial intelligence.

✎New Words

pave	[peɪv]	vt.铺设;为……铺平道路;安排
disseminate	[dɪˈsemɪneɪt]	vt.散布,传播
framing	[ˈfreɪmɪŋ]	n.构架,框架,骨架
immense	[ɪˈmens]	adj.极大的,巨大的
sphere	[sfɪə]	n.范围;势力范围
		vt.包围,围绕
embrace	[ɪmˈbreɪs]	v.拥抱;包括,包含;接受
self-assess	[self-əˈses]	n.自我评估,自主评估
self-decide	[self-dɪˈsaɪd]	n.自主决定
arena	[əˈriːnə]	n.表演场地,舞台
miracle	[ˈmɪrəkl]	n.奇迹,令人惊奇的人(或事)
empower	[ɪmˈpaʊə]	vt.授权;准许;使能够
thrill	[θrɪl]	vt.使兴奋,使激动
fusion	[ˈfjuːʒn]	n.融合
fantastic	[fænˈtæstɪk]	adj.极好的;很大的
symbiotic	[ˌsɪmbaɪˈɒtɪk]	adj.共生的
rigorous	[ˈrɪgərəs]	adj.严密的;缜密的;严格的
futuristic	[ˌfjuːtʃəˈrɪstɪk]	adj.未来的;未来派的;未来主义的
co-bot	[ˌkəʊ-bɒt]	n.协作机器人
updation	[ˈʌpdeɪʃn]	n.上升
upliftment	[ˈʌplɪftmənt]	n.提升
interference	[ˌɪntəˈfɪərəns]	n.干涉,干扰,冲突;妨碍
shortcoming	[ˈʃɔːtkʌmɪŋ]	n.短处,缺点
infuse	[ɪnˈfjuːz]	vt.灌输,使充满;鼓舞,激发
implication	[ˌɪmplɪˈkeɪʃn]	n.含义,含意
elaborate	[ɪˈlæbərət]	vi.详尽说明;变得复杂
		vt.详细制定;详尽阐述
surgery	[ˈsɜːdʒərɪ]	n.外科学,外科手术
app	[æp]	n.计算机应用程序
		abbr.应用(Application)
robotics	[rəʊˈbɒtɪks]	n.机器人技术
generic	[dʒəˈnerɪk]	adj.类的,属性的;一般的

✎Phrases

instant decision	即时决策
data storehouse	数据仓库
come up with	追赶上；提出；想出；设法拿出
nascent stage	初级阶段
data analytic	数据分析
situation oriented	情境导向
symbiotic relation	共生关系
human behavior	人类行为
a fraction of	一小部分
learning phase	学习阶段
self-learning mechanism	自学习机制
formulate business strategies	制定商业策略,制定业务策略
critical period	关键时期
business forecast	业务预测
intelligent cloud	智能云
smart data	智能数据
intelligent cloud computing	智能云计算
personal assistant	个人助手,个人助理

✎Abbreviations

IoT(Internet of Things)	物联网

✎Exercises

【Ex.5】 根据课文内容回答问题。

1. What has cloud computing always been?

2. What can machine learning and the cloud computing together do?

3. What are technicians are working towards?

4. What are the machine learning capabilities of these bots based on?

5. Why has the learning need for AI has increased in demand?

6. What can companies do?

7. What is a cloud going to become in future?

8. What does an intelligent cloud do in the educational sector?

9. In the scenario of an increasing paradigm shift from basic level machine learning to

the deep learning, what do AI-based cloud apps and computing algorithms do?

10. What do the Google Alexa and Microsoft Cortana have?

Reading

How AI and Big Data Are Connected?

Big data and AI are two of the most popular and useful technologies today. Artificial intelligence is in existence from more than a decade, while big data came into existence just a few years ago. Computers can be used to store millions of records and data, but the power to analyze this data is provided by the big data.

1. Big Data and AI

Big data and AI are considered two mechanical giants by data scientists, or other big corporations. Many organizations consider that AI will bring the revolution in their organizational data. Machine learning is considered as an advanced version of AI through which various machines can send or receive data and learn new concepts by analyzing the data. Big data helps the organizations in analyzing their existing data and in drawing meaningful insights[1] from the data.

Here, for example, we can imagine[2] a leather garment[3] manufacturer that exports its garments to the Europeans but he does not have any idea about the customer interests. Just by collecting data from the market and analyzing it through various algorithms, the merchant can identify the customer behavior[4] and interests[5]. According to their interests, they can provide the cloths. For this, the algorithms can help to find insight and accurate information too.

2. How Big Data Helps in AI Experiments

As is known that AI will reduce the overall human intervention[6] and jobs, people

---

① insight ['ɪnsaɪt] n.洞察力,洞悉;见解
② imagine [ɪ'mædʒɪn] vi.想象;猜想,推测
③ garment ['gɑːmənt] n.衣服,服装 v.给……穿衣服
④ behavior [bɪ'heɪvjə] n.行为,态度
⑤ interest ['ɪntrɪst] n.兴趣,爱好
⑥ intervention [ˌɪntə'venʃn] n.干涉,干预;介入

consider that AI has all machine learning capabilities and will create robots that will take over human jobs. The human role will be reduced due to AI expansion and this thought has been broken and changed by the involvement of the big data. Machines can take decisions on the basis of facts but cannot involve emotional interaction①, but due to big data the data scientists can involve their emotional intelligence and take the proper decisions in the right manner.

For a data scientist of any pharmaceutical② organization, he cannot only analyze the needs of customers but also inhibit③ the local rules and regulations of the particular market of that region. Depending on the salts used in any medicine, they can suggest the best options for that market, while in case of machine learning it may not be possible.

So, it is clear that the merge of AI and big data cannot only involve the talent and learning simultaneously, but also give rise to many new concepts and options for any new brand and organization. A mix④ of AI and big data can help the organizations to know the customer's interest in the best way. By using machine learning concepts, the organizations can identify the customer's interests in minimum possible time.

3. How Can Big Data Help in Global Diversification⑤?

With every passing day, the new technologies and tools are introduced in the market so the cost of tools of machine learning and AI are also reducing significantly. As a result of a price drop⑥, the technology will be adopted by a number of organizations. Even in a global region with diverse culture⑦, language, the technology and tools will be adopted with the same enthusiasm⑧. At the same time, the provider will have to provide the equivalent solutions to the market according to the customer needs.

Big data technology and tools will help the organizations in providing the relevant solutions to the customers as per their region and language, while at the same time machine learning will help them in providing the solutions to the organizations in the way so that customer sentiments will not get a heart. Like for any women-oriented

---

① emotional interaction：情感互动
② pharmaceutical [ˌfɑːməˈsuːtɪkl] adj.制药的，配药的 n.药物
③ inhibit [ɪnˈhɪbɪt] v.抑制；禁止
④ mix [mɪks] v.混合
⑤ diversification [daɪˌvɜːsɪfɪˈkeɪʃn] n.多样化；多种经营
⑥ price drop：价格下降
⑦ diverse culture：多元文化
⑧ enthusiasm [ɪnˈθjuːziæzəm] n.热情,热忱

product①, the way to market the product will be entirely different for Sri Lankan and Iranian markets as the sentiments of the women of both the regions may be entirely different.

4. Big Data and AI to Boost Market Analysis Insights

Right now, the market of big data and artificial intelligence is in their novice state and service providers do not have any idea where their customers are exactly and what are their needs. With time, they will realize the exact customer requirements and plan the offers and product functionalities② accordingly. With time, organizations will realize that what the exact needs of their customer requirements are. Even AI-based solutions may need to undergo huge changes as the customers' requirement may vary.

5. AI Technologies that Are Being Used with Big Data

There are several AI technologies that are used with big data and the following are a few of them.

5.1 Anomaly Detection③

For any dataset, if an anomaly is not detected then big data analytics can be used. Here fault detection④, sensor networks, eco-system distribution system⑤ health can be detected with big data technologies.

5.2 Bayes Theorem

Bayes theorem⑥ is used to identify the probability of an event based on the pre-known conditions⑦. Even the future of any event can also be predicted on the basis of the previous event. For big data analysis this theorem is of best use and can provide a likelihood of any customer interest in the product by using the past or historical data pattern.

---

① women-oriented product：面向妇女的产品
② functionality [ˌfʌŋkʃəˈnælətɪ] n. 功能，功能性
③ anomaly detection：异常检测
④ fault detection：故障检测
⑤ eco-system：生态系统
⑥ Bayes theorem：贝叶斯定理
⑦ pre-known condition：已知条件

5.3　Pattern Recognition

Pattern recognition① is a technique of machine learning and is used to identify the patterns in a certain amount of data. With the help of training data, the patterns can be identified and are known as supervised learning.

5.4　Graph Theory

Graph theory② is based on graph study that uses various vertices③ and edges. Through node relationships, the data pattern and relationship can be identified. This pattern can be useful and help the big data analysts in pattern identification④. This study can be important and useful for any business.

6．Summary

It can be said clearly that AI and big data are two of the emerging technologies that are used by organizations extensively. The technologies are used to provide better customer experience⑤ in an organized and smarter way. They can be blended to provide a seamless⑥ experience to customers.

AI and big data use many methods and techniques, but they can be used in an integrated manner and provide a result to be used by the organizations to analyze customer interests and offer them the best-optimized services.

参考译文

人工智能在不久的未来需求旺盛

　　2018 年,人工智能平台和应用程序大幅增加。随着软件和互联网行业的发展,该技术风靡医疗保健、法律、制造业、汽车和农业等其他行业。早些时候,人工智能仅限于科幻电影,但今天,我们可以看到技术已经赶上了想象力,梦想成真。人工智能已成为现实。我们遇到的每个人都习惯于在他或她的日常生活中使用人工智能。一个专家小组报告说,人工智能的兴起是一场工业革命,与蒸汽机、石油和电力以及计算机的三次工业革命

---

① pattern recognition：模式识别
② graph theory：图论
③ vertex ['vɜːteks] *n*. 顶点
④ pattern identification：模式识别
⑤ customer experience：客户体验
⑥ seamless ['siːmlɪs] *adj*. 无缝的

相提并论。

很快我们就会发现人工智能革命成为门户,它成为将城市变为更广泛"数字革命"中的"信息基础设施"。最近的一份报告称,23%的企业已将该技术纳入业务流程。市场研究称,到 2025 年,全球人工智能软件市场预计将增至 1186 亿。

根据最近的一项调查显示,人工智能的重大进展和改变游戏规则仅需几个月的时间。AI 无疑将成为不久未来的商业和社会的驱动力。

让我们看看 AI 在不久的未来会取得的进展。

1. 聊天机器人

处理自然语言的技巧被广泛认为是智力的标志。人工智能的发展使聊天机器人取代了传统的会话服务。由于聊天机器人是根据过去的信息进行训练的,因此组织会使用它们来维护报告。该技术利用这些日志来想出客户要查询的问题。通过融合会话历史和机器学习工具,将给客户适当的答案。

聊天机器人有以下三种模式。

1.1 模式匹配

该技术使用人工智能标记语言对文本进行分组并产生适当的响应。聊天机器人尝试将查询与数据库中存在的模式组合相关联。

1.2 自然语言理解

自然语言理解算法分析人类查询,这不会受到过去会话的影响。为了区分早期的对话,其状态被存储。它可能是在餐馆、办公室等地。这些参数更有助于将查询与上下文联系起来,而不必担心聊天的历史记录。

1.3 自然语言处理

自然语言处理把文本转换为结构化数据。它组合使用以下给定步骤:

(1)标记化:句子或单词系列被分为具有语言代表性的标记或片段。

(2)情感分析:自然语言处理研究人类体验,为人类提供满意的答复。

(3)规范化:处理文本以检测那些改变了用户查询含义的印刷错误和拼写错误。

(4)命名实体识别:程序模型搜索与查询相关的特定单词(如姓名、地址等)。

(5)依存句法分析:它在用户的文本中查找主题、动词、常用短语、名词和对象,以找出与用户请求相关的短语。

人工智能驱动的聊天机器人可以轻松处理复杂的人机交互。企业已采用这种支持人工智能的聊天机器人为客户提供个性化体验。聊天机器人旨在处理人工查询并轻松指导客户完成复杂的程序。

2.　虚拟助理

近十年以来,数字环境目前正在出现一种趋势——承认虚拟助理。近年来,人们日益接受了像 Siri、Alexa、Google 助手等虚拟助理。全球市场情景表明,组织对多任务处理能力的需求日益增长,这使得人们接受了虚拟助理。据估计,到 2021 年,全球使用虚拟助手的人数预计将达到 18 亿。大多数智能手机上的这些虚拟助理都能理解用户的语音命令并按要求执行任务。它们是芯片、麦克风和软件的组合,可识别语音查询并使用你选择的语音进行响应。

虚拟助理在商业环境中找到了新的用途。一项调查显示,组织正在使用虚拟助手来提高效率。以下是虚拟助理在企业中的用途。

2.1　自动协调会议

如果虚拟助理可以访问所有员工和客户的组织日历,则可以根据每个人的日程安排预订会议时段。如果会议被取消,一些虚拟助理会通知外部参会者。

2.2　联系客户

虚拟助理使组织能够直接与客户进行交互。本地化操作可帮助公司根据位置、语言和区域自定义内容和体验。此外,它们还有助于开发兼容的应用程序,可以收集客户数据以改善销售和推广。

2.3　管理基础设施

语音助理能够监视组织的后端计算服务。他们可以监控可用资源,刺激安全警报和执行的任务。

2.4　启用智能办公室

语音助理使智能办公室的概念更易于管理。数字助理可以根据应用情况来控制房间内的灯光、投影仪等设施。它们还会在资源不可用时报警。

虚拟助理使人的生活更加舒适。我们已经开始依靠语音助理来播放广播,接收天气预报,阅读电子邮件,打电话和接收交通新闻等。

3.　面部识别

数字传感器、处理能力、数据分析和神经网络的发展使面部识别系统具有惊人的精确度。面部识别系统使用 LBPH 算法来检测人的眼睛、鼻子、嘴唇、耳朵和下巴的轮廓。面部识别算法甚至可以设置地能够判断情绪。

面部识别系统用以下两种模式运行。

3.1   面部图像的验证或认证

在该模式中,把输入面部图像与认证所需的图像进行比较。通常它是一对一相关的。

3.2   识别或面部识别

将输入的面部图像与数据集中的所有图像进行比较以找到匹配。这种比较是一对多的。

4. 人工智能芯片

深度学习和机器学习模型越来越被接受,因此需要使用功能强大的芯片来处理大量数据。过去几年的技术进步已经推动人工智能芯片取得了比以往更多的胜利。接下来的几年将会看到专用芯片的兴起,这些芯片的制造是为了执行复杂的数学计算并提升人工智能应用程序的执行速度。这些人工智能芯片将加快对象检测和面部识别等任务的执行速度。智能手机的人工智能芯片可提高数据保密性和安全性。

这些芯片针对计算机视觉、自然语言处理和语音识别的特定用例和场景进行了优化。下一代产业将依靠这些芯片为最终用户提供智能。

5. 结论

在不久的将来,人工智能肯定会在市场上引起轰动。人工智能的进步不会放缓。人工智能不仅被软件和互联网行业所接受,而且在医疗保健、汽车、零售等领域也越来越受欢迎。它让商业运营和人类生活更轻松。一个更光明的未来等待着每个人,使用人工智能,人类不再做烦琐的工作。

Unit 10

录音

Text A

Artificial Intelligence: Our Future Computing

Today, our morning routines will probably see us checking our smartphones for updates on the latest news headlines and updates of our friends' social lives. You read the work emails arrived overnight, text your sister to confirm dinner plans, and perhaps check traffic conditions as you head to work.

In 2038, or probably earlier, digital devices will help us do more with one of our most precious commodities: time. At Microsoft, we imagine a world where your personal digital assistants such as Microsoft's Cortana will be trained to anticipate our needs, help manage our schedule, prepare us for meetings, assist as we plan our social lives, reply to and route communications, and drive cars.

Artificial intelligence will enable breakthrough advances in areas like healthcare, agriculture, education and transportation. It is already happening in impressive ways.

In the office, AI is already helping companies understand customers better by analysing their preference and consumption patterns. In the fields, AI is helping farmers analyse weather patterns and soil conditions to help optimise their harvest. At home, smart devices such as lighting control and music players work together to provide the perfect ambience to unwind after a long day.

Beyond becoming the catalyst for some of the most dramatic changes in the way we live, work and play today, embracing AI will yield both economic and societal benefits.

By 2021, digital transformation will add an estimated $10 billion to Singapore's gross domestic product (GDP), according to a study released last month by research firm IDC, in partnership with Microsoft. The study also found that digital transformation will boost growth by an additional 0.6 per cent annually. This will accelerate as digital technologies and services are enabled by AI, along with the Internet of Things (IoT) and data analytics.

The projected economic growth is possible with businesses adopting transformative technologies, of which AI is a central piece. In this future economy, new startups will sprout even as traditional businesses find new niches to grow into.

But as we have witnessed over the past 20 years, new technology also inevitably raises complex questions and broad societal concerns.

How do we ensure that AI is designed and used responsibly? How do we establish ethical principles to protect people? How should we govern its use? And how will AI impact employment and jobs?

To answer these tough questions, technologists will need to work closely with government, academia, business, civil society and other stakeholders. At Microsoft, we identified six ethical principles-fairness, reliability and safety, privacy and security, inclusivity, transparency, and accountability-to guide the cross-disciplinary development and use of artificial intelligence. The better we understand these or similar issues—and the more technology developers and users can share best practices to address them—the better served the world will be as we contemplate societal rules to govern AI.

We must also pay attention to AI's impact on workers. One constant technological change of the past 250 years has been the ongoing impact of technology on jobs—the creation of new jobs, the elimination of existing jobs and the evolution of job tasks and content.

Some key conclusions are emerging.

First, the companies and countries that will fare best in the AI era will be those that embrace these changes rapidly and effectively. This is because new jobs and economic growth will come to those that embrace the technology, not those that resist or delay adopting it.

Second, while we believe that AI will help solve big societal problems, we must be critical. There will be challenges as well as opportunities. We must address the need for strong ethical principles, the evolution of laws, training for new skills and even labour market reforms. This must all come together if we are going to make the most of AI.

Third，we need to act with a sense of shared responsibility because AI will not be created by the tech sector alone. At Microsoft，we are working to democratise AI in a manner that is similar to how we made the PC available to everyone. This means we are creating tools to make it easy for every developer，business and government to build AI-based solutions and accelerate the benefit to society.

All of this leads us to what may be one of the most important conclusions of all. Skilling-up for an AI-powered world involves more than science，technology， engineering and math. As computers behave more like humans，the social sciences and humanities will become even more important. Languages，art，history，economics， ethics，philosophy，psychology and human development courses can teach critical， philosophical and ethics-based skills that will be instrumental in the development and management of AI solutions. If AI is to reach its potential in serving humans，then every engineer will need to learn more about the liberal arts and every liberal art major will need to learn more about engineering.

Singapore's ability to nurture a future-ready workforce will be key to thriving in the digital economy，one where digital products and services will make up 60 per cent of our economy. We need to constantly reassess and prepare ourselves for a future that appears to be shaping up differently each time you take another look at the crystal ball.

Yet，that concept is not new to Singapore，a young country that has both experienced and embraced change over the years. Singapore's success has always depended on the skill，intelligence and determination of its people. Aided by AI，this human ingenuity will be even more important in the years ahead.

With AI，we have more power at our fingertips than entire generations that came before us. The questions will be what we will do with it.

✎ New Words

routine	[ruːˈtiːn]	n.常规，例行程序
		adj.例行的，常规的，日常的
headline	[ˈhedlaɪn]	n.大字标题；新闻提要；头条新闻
precious	[ˈpreʃəs]	adj.珍贵的，贵重的
commodity	[kəˈmɒdətɪ]	n.商品，日用品
anticipate	[ænˈtɪsɪpeɪt]	vt.预感，预见，预料
schedule	[ˈʃedjuːl]	n.进度表，明细表；预定计划
		vt.排定，安排
breakthrough	[ˈbreɪkθruː]	n.突破；重要技术成就

impressive	[ɪmˈpresɪv]	adj.给人印象深刻的,引人注目的;可观的
preference	[ˈprefrəns]	n.优先权;偏爱
consumption	[kənˈsʌmpʃn]	n.消费
optimise	[ˈɒptɪmaɪz]	vt.使最优化
harvest	[ˈhɑːvɪst]	n.收成;结果
		v.收割,收成
ambience	[ˈæmbɪəns]	n.气氛,周围环境
catalyst	[ˈkætəlɪst]	n.催化剂,促进因素
firm	[fɜːm]	n.公司,企业
		adj.坚固的,确定的
transformative	[ˌtrænsˈfɔːmətɪv]	adj.有改革能力的,变化的,变形的
startup	[ˈstɑːtʌp]	n.新兴公司;启动
sprout	[spraʊt]	vi.发芽;抽芽
		vt.使发芽;使生长
niche	[nɪtʃ]	n.商机
inevitably	[ɪnˈevɪtəblɪ]	adv.必然地,无疑地
responsibly	[rɪˈspɒnsəblɪ]	adv.负责地,有责任感地
principle	[ˈprɪnsɪpl]	n.原则,原理
academia	[ˌækəˈdiːmɪə]	n.学术界,学术环境
stakeholder	[ˈsteɪkhəʊldə]	n.股东;利益相关者
fairness	[ˈfeənɪs]	n.公正,公平
inclusivity	[ˌɪnkluːˈsɪvɪtɪ]	n.包容性
transparency	[trænsˈpærənsɪ]	n.透明度,透明性
accountability	[əˌkaʊntəˈbɪlɪtɪ]	n.有责任,有义务,责任制
cross-disciplinary	[krɒs-ˈdɪsəplɪnərɪ]	n.跨学科
contemplate	[ˈkɒntəmpleɪt]	vt.周密考虑
		vi.沉思,深思熟虑
resist	[rɪˈzɪst]	v.抵抗,抗拒,抵制
evolution	[ˌiːvəˈluːʃn]	n.演变,进化,发展
democratise	[dɪˈmɒkrətaɪz]	v.民主化
developer	[dɪˈveləpə]	n.开发者
humanity	[hjuːˈmænətɪ]	n.人文学科
ethic	[ˈeθɪk]	n.伦理,道德
philosophy	[fɪˈlɒsəfɪ]	n.哲学
instrumental	[ˌɪnstrəˈmentl]	adj.有帮助的,起作用的
workforce	[ˈwɜːkfɔːs]	n.劳动力,劳动人口
thrive	[θraɪv]	vi.兴盛,茁壮成长
reassess	[ˌriːəˈses]	v.再估价,再评价

| fingertip | [ˈfɪŋɡətɪp] | *n*.指尖 |

Phrases

consumption pattern	消费模式
soil condition	土壤情况
societal benefit	社会效益
societal concern	社会关注
societal rule	社会规则
economic growth	经济增长
ethical principle	道德原则
labour market	劳动力市场
tech sector	技术部门
liberal art	文科
be shaping up	正在成形
crystal ball	（占卜用的）水晶球，预言未来的方法

Abbreviations

| GDP（Gross Domestic Product） | 国内生产总值 |

Exercises

【Ex. 1】 根据课文内容回答问题。

1. What will digital devices do in 2038, or probably earlier?

2. In what areas will artificial intelligence enable breakthrough advances?

3. What is AI doing in the fields?

4. By 2021, how much will digital transformation add to Singapore's gross domestic product（GDP）, according to a study released last month by research firm IDC, in partnership with Microsoft?

5. What are the six ethical principles identified to guide the cross-disciplinary development and use of artificial intelligence?

6. What has one constant technological change of the past 250 years been?

7. What will be the companies and countries that will fare best in the AI era? Why?

8. Why do we need to act with a sense of shared responsibility?

9. What will every engineer need to do if AI is to reach its potential in serving humans?

10. What has Singapore's success always depended on?

【Ex.2】 把下列单词或词组中英互译。

1. ethical principle   1. _____
2. anticipate   2. _____
3. cross-disciplinary   3. _____
4. preference   4. _____
5. optimise   5. _____
6. *n*.进度表,明细表;预定计划   6. _____
7. *n*.透明度,透明性   7. _____
8. *n*.常规,例行程序   8. _____
9. *n*.伦理,道德   9. _____
10. *n*.新兴公司;启动   10. _____

【Ex.3】 短文翻译。

China Expected to Lead AI Application in Smart City Development

China's rapid deployment of new technologies and desire to develop artificial intelligence (AI) will help lead the smart city development, said experts on Nov 15.

Senior information technology experts and entrepreneurs in Silicon Valley shared their opinions on China's smart city development and application at a seminar held by the US-Asia Technology Management Center and Department of East Asian Languages and Cultures at Stanford University in Palo Alto, California.

"China started implementing smart cities many years ago and it has developed very quickly, because Chinese people are enthusiastic about adopting new technologies," said Yu Wenli, business executive of technology companies in Silicon Valley.

In China, sensors were installed almost everywhere, such as street sensors for traffic tracking and control, and garage sensors for guiding drivers to find the nearest empty parking space, he said.

China started piloting national smart city development in 2012 to encourage the use of new technologies, such as AI and Internet of Things (IoT), to help traffic flows, improve law enforcement and make public buildings more energy efficient.

'Almost every city's mayor is talking about smart city, big data, AI and cloud computing,' said Li Yangming, a senior IT expert at China National Petroleum Corporation and a visiting fellow at the Stanford Asia-Pacific Research Center.

'Currently, there are about 10 Alibaba City Brain projects operating by using AI algorithms to monitor and control traffic signals and street cameras,' he said.

"The United States and China are the world's leaders in developing AI and smart city technologies. While the United States leads in technological innovation, China enjoys advantages in data collection, investment in infrastructure and government support," Yu said.

【Ex.4】 将下列词填入适当的位置（每个词只用一次）。

information	understand	streamline	smart	communicate
embedded	improve	shift	validate	behavior

Making Cities Smarter with AI

1. A bot to answer all questions

Organizations use chatbots for initial customer contact and support purposes. This concept can be adopted by government CIOs as AI-powered chatbots can contextualize and personalize government services, __1__ service delivery and augment municipal employees' effectiveness.

Well-designed conversational platforms __2__ the burden of dealing with complexity from the users to the technology. Computers have to __3__ humans, not the other way around.

2. Process the normal, detect the abnormal

AI excels in processing routine requests and detecting unusual __4__. Government CIOs should exploit these capabilities to simplify and __5__ their services. CIOs should begin using AI and conversational chatbots as one way to __6__ and disseminate information on anomalies or abnormalities while they deal with citizen or user requests. AI technology can detect patterns in claims that could signal fraud, as well as simple user errors.

Advanced analytics and data science, including machine learning algorithms, can cross-reference data and __7__ complex business or service processes. These outcomes, accessed by citizens via AI chatbots, are a key benefit in __8__ city services.

Although AI can help drive these services, embedding AI in smart city solutions requires changes in city operations, IT platforms and data privacy policies. While an __9__ and interoperable AI improves the range of applications, the complexity of __10__ and data flow increases and raises new questions on algorithmic business flows. For example, if a technology provider develops an AI chatbot, who owns the intellectual property it generates?

Text B

Artificial Intelligence in Smart Cities

Artificial intelligence (AI) could play a key role in numerous applications within a smart city, from improvements to traffic and parking management to the safe integration of autonomous ride-share vehicles.

However, in many cases, human creativity is necessary in order to teach an AI programme the models and patterns that need to be recognised. Once trained, this AI can then sift through a monumental volume of data in a flash, and at high accuracy.

"Basic AI, or machine learning to be more apt, is ideal for adoption in the smart city," said Randi Barshack, Chief Marketing Officer of Figure Eight, a San Francisco-headquartered AI platform, "There are many layers that can be unpeeled in terms of possibilities."

1. Smile, you're on camera

A significant application for AI in the smart city is video surveillance, and in some cities today, closed-circuit television (CCTV) cameras already use AI for facial recognition. In December 2017, a BBC reporter demonstrated how this technology could be used for security purposes, and was tracked down by AI in the Chinese city of Guiyang in less than ten minutes. In Zhengzhou, police officers are using 'smart' AI glasses to the same effect, recognising criminal suspects and finding civilians with fake IDs.

AI could be used not only to apprehend, but also to provide assistance. If a pedestrian collapses in the street, camera-based AI could detect that a medical professional is required on the scene. Facial recognition technology could also source the person's name, age and home address, and pass this information on to the relevant authorities. "But it's not just that this technology can recognise a face or an anomaly, it can also register facial expressions and gesture recognition," added Barshack, "The number of things you can do with camera-based AI is virtually endless, but it is critical for municipalities and cities to first understand the problems they need to solve, and how AI can be applied to solve them."

For example, if traffic congestion is an issue, there are a variety of solutions such as adjusting the way traffic lights are metered, or building or closing new roads. AI can be used to process this traffic data, but there must be an initial theory for the algorithm to work with in the first place. This is where human creativity is necessary. As Barshack explained, "AI can do in milliseconds what it might take years for a human to process, but you need to have those data points, and you need to have the theory. AI is not at the point where you can just say: 'make the traffic patterns better'."

By using human intelligence to create potential solutions, AI can then be used to model the results of proposed measures. This also allows for corner cases to be better understood beforehand. "Those 'what if' scenarios can be explored at speeds you never would have imagined before," she explained, "What might have taken years to prove out can now be done in a matter of minutes."

2. Future mobility brings new challenges

A pillar to any smart city is the provision of accessible, affordable and clean mobility services—be it public transport or through private vendors that offer ride-sharing, bike sharing and on-demand vehicle hire. In many cities, ride-sharing has become particularly popular, but there have been teething pains for city planners.

The nature of ride-sharing results in these vehicles making numerous stops on a single journey, much like a city bus. However, a lack of dedicated pick-up and drop-off areas can lead to a spike in congestion and an increased likelihood of a road traffic incident—the antithesis of a smart city road network. "Traffic is a significant issue in San Francisco these days, for example, and much of that traffic occurs because there are no spots where ride-sharing vehicles can pull over," said Barshack. "If spaces were freed up, what would happen to parking congestion?" Human-in-the-loop AI could help city planners to understand this, she suggested.

Autonomous cars are also considered a significant element to any smart city. Today, the vast majority of a vehicle's time is wasted as it is either stationary in a parking lot or at home. This creates a situation where many drivers are simply searching for a parking space, exacerbating congestion issues. "In theory, these vehicles could be moving, transporting and managing our lives in a much more interesting way," said Barshack, pointing to the opportunities of a fully autonomous vehicle.

Over the past decade or so, autonomous test vehicles have been taught how to recognise a plethora of objects they may encounter on the road and surrounding environment. Anything from a lamppost or pedestrian, to a dog or parked truck has been programmed into the system, with an understanding of how these actors may walk, run, change lanes or pull into the road. However, in a smart city, there will not only be cars and pedestrians to worry about.

For example, one of the fastest growing trends in Chinese cities is bicycle sharing, while in California, electric scooter hire has become increasingly popular. With improved air quality and greater ease of access to mobility in mind, city planners are encouraging such alternatives to private vehicle ownership. This will pose a number of new challenges to autonomous driving AI in coming years, which will need to become increasingly capable of driving amid these new parties.

"In San Francisco, e-scooters have suddenly popped up all over the place," said Barshack, "but a person on a scooter could be really confusing for an autonomous car. It thinks it sees a pedestrian, but its trajectory is faster than the average speed of a pedestrian." In April 2018, San Francisco city authorities blasted a number of electric scooter-share companies for deploying without permission. Proposed legislation would see all operators require a permit, with penalties for scooters that are not 'parked responsibly'. Bikes and baby strollers can also pose issues to autonomous driving AI.

"Development teams will be ensuring that autonomous driving models understand all these nuances, such as the difference between a pedestrian and a person on a scooter," continued Barshack. "Policy and lawmakers also need to be educated on what is happening in this space; if you're used to regulating horses, and cars suddenly come in, you need to understand what you're dealing with."

Machine learning needs human creativity. It is clear that AI will have a part to play in the smart city, both in terms of improving the capability of autonomous vehicles and in assisting the development of roadways that can handle new forms of mobility. Citizens can also expect to be under more accurate surveillance to reduce crime and improve incident response efficiency.

City authorities will be able to work through huge quantities of data to test and deploy new initiatives to cope with demand for parking, and access to ride-share vehicles. What is interesting is that despite the push to develop AI that is more capable than a human, it is the human element that will prove vital in training these algorithms in the first place.

⚓ New Words

traffic	['træfɪk]	n. 交通,运输量
creativity	[ˌkriːeɪ'tɪvɪtɪ]	n. 创造性,创造力
recognise	['rekəgnaɪz]	vt. 认出,识别出某人[某事物];认可,承认
monumental	[ˌmɒnjʊ'mentl]	adj. 重要的;非常大的
apt	[æpt]	adj. 恰当的,适当的;聪明的,灵巧的
unpeel	['ʌn'piːl]	v. 削……的皮,剥离
possibility	[ˌpɒsə'bɪlɪtɪ]	n. 可能,可能性

camera	['kæmərə]	n.摄像头,摄影机;照相机
demonstrate	['demənstreɪt]	vt.证明,证实;显示,展示
security	[sɪ'kjʊərɪtɪ]	n.安全;保证,保护
		adj.安全的,保安的,保密的
apprehend	[ˌæprɪ'hend]	vt.逮捕,拘押
collapse	[kə'læps]	vi.(尤指工作劳累后)坐下
anomaly	[ə'nɒməli]	n.异常,反常
register	['redʒɪstə]	n.&vt.登记,注册
congestion	[kən'dʒestʃən]	n.拥挤,堵车
millisecond	['mɪlɪsekənd]	n.毫秒
measure	['meʒə]	n.测量;措施;程度
		vt.测量,估量
beforehand	[bɪ'fɔːhænd]	adv.事先,预先;提前
		adj.提前的;预先准备好的
imagine	[ɪ'mædʒɪn]	vt.设想;想象;猜想;误认为
		vi.想象;猜想,推测
pillar	['pɪlə]	n.台柱,顶梁柱
provision	[prə'vɪʒn]	n.设备;供应
affordable	[ə'fɔːdəbl]	adj.付得起的
on-demand	[ˌɒndɪ'mɑːnd]	adj.按需的
dedicated	['dedɪkeɪtɪd]	adj.专注的,专用的
pick-up	['pɪkʌp]	n.提取
drop-off	['drɒpɒf]	n.急下降,直下降
antithesis	[æn'tɪθəsɪs]	n.对立,对立面
stationary	['steɪʃənrɪ]	adj.不动的,固定的;静止的,不变的
exacerbate	[ɪg'zæsəbeɪt]	vt.使恶化;使加重
plethora	['pleθərə]	n.过多,过剩
surround	[sə'raʊnd]	vt.包围,围绕
environment	[ɪn'vaɪrənmənt]	n.环境,外界
encourage	[ɪn'kʌrɪdʒ]	vt.鼓励,支持
suddenly	['sʌdənlɪ]	adv.意外地,忽然地
trajectory	[trə'dʒektərɪ]	n.轨道
blast	[blɑːst]	vi.严厉批评或猛烈攻击
permission	[pə'mɪʃn]	n.允许,批准,认可
propose	[prə'pəʊz]	vt.提议,建议;打算,计划
		vi.做出计划,打算
legislation	[ˌledʒɪs'leɪʃn]	n.立法;法律,法规
penalty	['penəltɪ]	n.惩罚,刑罚

nuance	['njuːɑːns]	n. 细微差别
roadway	['rəʊdweɪ]	n. 路面, 道路; 车道
mobility	[məʊ'bɪlɪtɪ]	n. 流动性; 移动性; 机动性
initiative	[ɪ'nɪʃətɪv]	n. 主动性; 主动权
		adj. 自发的; 创始的

✎ Phrases

smart city	智慧城市
parking management	存车管理, 停车管理
ride-share vehicle	共享汽车, 拼车
in a flash	立刻, 一瞬间
video surveillance	视频监控
traffic congestion	交通拥堵
public transport	公共交通, 公共交通工具
city planner	城市规划师
a lack of	缺乏, 缺少
traffic incident	交通事故, 交通事件
freed up	被释放
pull into	开进, 开向路边
bicycle sharing	共享自行车
electric scooter	电动车, 电动踏板车, 电动代步车
autonomous driving	自主驾驶
baby stroller	婴儿车
be educated on	接受教育

✎ Abbreviations

CCTV（Closed-Circuit TeleVision） 闭路电视

✎ Exercises

【Ex.5】 根据课文内容填空。

1. However，in many cases，_____ is necessary in order to teach an AI programme the models and patterns that need to _____. Once trained, this AI can then sift through a monumental volume of data _____, and _____.

2. A significant application for AI in the smart city is _____, and in some cities today，closed-circuit television（CCTV）cameras already use AI for _____.

3. In Zhengzhou，police officers are using _____ to the same effect，recognising criminal suspects and _____ .

4. AI could be used not only to apprehend，but also to _____ . If a pedestrian collapses in the street，camera-based AI could detect that _____ is required on the scene. Facial recognition technology could also source the person's _____ , _____ and _____ , and pass this information on to the relevant authorities.

5. By using human intelligence to create potential solutions，AI can then be used to _____ . This also allows for corner cases _____ beforehand.

6. A pillar to any smart city is the provision of _____ , _____ and _____ —be it public transport or through private vendors that offer ride-sharing， _____ and on-demand vehicle hire.

7. A lack of dedicated pick-up and drop-off areas can lead to _____ and an increased likelihood of a road traffic incident—the antithesis of _____ .

8. Over the past decade or so，autonomous test vehicles have been taught how to _____ they may encounter on the road and _____ .

9. _____ needs human creativity. It is clear that AI will have a part to play in the smart city，both in terms of improving the capability of _____ and in assisting _____ the development of roadways that can _____ .

10. What is interesting is that despite the push to develop AI that is _____ than a human，it is the _____ that will prove vital in training these algorithms _____ .

Reading

The Future of Artificial Intelligence

Believe it or not，we all are leveraging[①] artificial intelligence like technology in our everyday life. Whether it is about locating the nearby barber shop to a simple order we placed over e-commerce marketplace，AI is somehow involved in it. A report by Statista reveals that the overall AI market will reach 7,35 billion USD by the end of this year. This market will expand up to 89,847.35 million USD by the end of 2025. For those who are willing to grow their career[②] with this advancement can participate in artificial intelligence course and various other resources like blogs，videos，coding platforms etc.，accessible over the internet. Here are the upcoming[③] AI trends we can

---

① leverage ['li:vərɪdʒ] v. 影响

② career [kə'rɪə] n. 职业，事业

③ upcoming ['ʌpkʌmɪŋ] adj. 即将来到的，即将出现的

see in the following years.

1. Automation Fulfilling Disaster Management Demands

Whether it is about chatbots automating the whole customer support or sensors generating millions of dataset covering temperature, pressure and many more attributes, AI is everywhere. AI experts are pushing their efforts to make machines more responsive[①] and attachable[②] to a human heart. The branch of AI which has been predicted with the highest advantage in machine intelligence is robotics. The powerful robots with complex algorithms can assist human in several ways. All the challenging roads on the human map can be chased[③] by robots. Complex predictions like weather forecasting or any disaster[④] events can be easily recognized and using precautions[⑤] with proper risk management[⑥] strategy will be available to deal with such situations.

These robots can deal with the most dangerous[⑦] situations which are beyond human capability. Robots can perform better in managing city traffic and disaster. Our modern analytics can detect hazardous[⑧] events and their impact to rescue[⑨] humans. With rescue apps, we can reach out to whole civilians delivering precautionary messages.

2. Autonomous Vehicles: A Hard Real-Time System

We have already encountered autonomous vehicles such as Google's Wyamo and Teslas cars. These self-driving vehicles have emerged as a power source[⑩] to eliminate the complexities played on a road through human's fault. Reduction of events like accidents, late delivery, lost and many more is the biggest priority of such vehicles. With self-driving cars achieving complete safety of a person has become possible. Also, its involvement in delivery, military and many more areas has provided its complete potential. We have achieved so much, but this technology is not yet so perfect. It demands improvement so that our whole generation can adopt it. Apart

---

① responsive [rɪˈspɒnsɪv] *adj.* 应答的，响应的；反应灵敏的
② attachable [əˈtætʃəbl] *adj.* 可附上的，可连接的
③ chase [tʃeɪs] *vt.* 追寻
④ disaster [dɪˈzɑːstə] *n.* 灾难
⑤ precaution [prɪˈkɔːʃn] *n.* 预防，防备
⑥ risk management：风险管理
⑦ dangerous [ˈdeɪndʒərəs] *adj.* 危险的
⑧ hazardous [ˈhæzədəs] *adj.* 冒险的，有危险的 *adv.* 冒险地，有危险地 *n.* 冒险，危险
⑨ rescue [ˈreskjuː] *vt.* 营救，救援
⑩ power source：电源，能源

from vehicles, the public transportation sector including buses and trains will also fall under this technology pretty soon.

3. Gaming Sector Will See Its Complete Potential

You must be thinking: What's the role of AI in gaming sector? Well, you must remember that last FIFA where Germany marked its powerful victory holding the world of a championship①. In that match, Germans leveraged the AI and other analytics resources to understand their opponents move. They crunched millions of historical and real-time gaming data② to defeat opponents③.

When it comes to the decision making, AI is the best option which uses the power of trained models eating qualified data. Once analyzed, it will provide you with the most valuable④ insights from the game making your next move very clear. Currently, researchers and Chelsea FC are participating together to understand a players decision. Here, they will understand what might have happened if a player has done that in order to make quick and accurate decisions.

4. Better Designs

When it comes down to innovation, designing new components is a very typical phase. Artificial intelligence can provide millions of new shapes with qualified configurations in just a few seconds. Big giants like General Motors, Airbus are participating in this fourth industrial revolution to bring pace in humanity. They are using complex AI algorithms to design their parts and many more sensitive⑤ components.

5. Assistance

Whether it is Apple's Siri or Amazon Alexa, they are developed to provide a personalized virtual assistance to you. These smart applications are the outcome of AI and expanding their legs across multiple zones where human require assistance. Currently, whenever you look for the best restaurant near you, all it requires is just a command. Similarly, the machine learning powered automated chatbots are emerging

---

① championship [ˈtʃæmpɪənʃɪp] n.锦标赛；锦标,优胜,冠军称号
② real-time gaming data：实时游戏数据
③ opponent [əˈpəʊnənt] n.对手 adj.对立的；敌对的
④ valuable [ˈvæljuəbl] adj.宝贵的,有价值的
⑤ sensitive [ˈsensɪtɪv] adj.敏感的,灵敏的

as a powerful resource to interact with a human through texts or voices. These talkative machines are just not limited to provide customer support, they also inherit critical algorithms① for the recommendation②, pattern recognition or self-improvement③, giving a personalized experience to the user. It has been predicted that in our upcoming time, we don't need to surf websites. The chatbots will be there to accomplish all our tasks. Websites, pizza delivery companies, banks and many other sectors are working hard to get more from this technology.

6. Security

When there is a rule, there are flaws. When AI was introduced to the common forgery and frauds of the real-world such as in loans, courtroom decision, cyber breach, fraud transactions etc., it identifies their patterns and comes up with a new solution. Yes, our FinTech sector is already into it, and at the same time, cyber security, laws, and many other sectors are looking for a cure for the holes enlarging in their services. The intrusion detection system④, the block chain⑤, decision making and many other processes are enhancing giving us accurate results. In our upcoming years, as more and more devices are getting connected over the internet each day, the circumstances for cyber attacks⑥ are also increasing. They all require security and AI can give them that.

参考译文

人工智能：我们的未来计算

如今，每天早上我们可能都会查看智能手机，了解最新的新闻标题以及更新的朋友圈。你阅读夜间收到的工作电子邮件，给你姐姐发短信确认晚餐计划，或者在上班的时候查看交通状况。

到 2038 年或者可能更早，数字设备将帮助我们更好地利用我们最贵的东西——时间。在微软，我们可以想象这样一个世界：你的个人数字助理（如微软的 Cortana）将接受训练，以预测我们的需求，帮助我们管理日程安排，准备会议，协助计划社交生活，回复和

① critical algorithm：关键算法
② recommendation [ˌrekəmenˈdeɪʃn] n. 推荐
③ self-improvement [self-ɪmˈpruːvmənt] n. 自我改善
④ intrusion detection system：入侵检测系统
⑤ block chain：区块链
⑥ cyber attack：网络攻击

发送信息以及驾驶汽车。

人工智能将在医疗保健、农业、教育和交通等领域取得突破性进展。现在的进展已经给人们留下了深刻的印象。

在办公室,人工智能已经能通过分析客户的偏好和消费模式,帮助公司更好地了解他们。在田间,人工智能正在帮助农民分析天气模式和土壤条件,以使得他们获得丰收。在家中,照明控制和音乐播放器等智能设备协同工作,为你在漫长的一天工作后提供完美的氛围。

人工智能成为我们今天生活、工作和娱乐方式中一些最显著变化的催化剂,除此之外,人工智能也将带来经济和社会效益。

研究公司 IDC 上个月与微软合作发布的一项研究表明,到 2021 年,数字化转型将为新加坡的国内生产总值(GDP)增加约 100 亿美元。该研究还发现,数字化转型将使年增长率再提高 0.6%。随着人工智能所提供的数字技术和服务以及物联网(IoT)和数据分析的出现,这个增长将会加速。

企业采用变革性技术可以实现预计的经济增长,人工智能是其中的核心。在未来的经济中,传统企业会发现新商机,新的创业公司也会开始萌芽。

但正如我们在过去 20 年中目睹的那样,新技术也不可避免地引发了复杂的问题和广泛的社会问题。

我们如何确保人工智能的设计和使用是负责任的? 我们如何制定道德原则来保护人民? 我们该如何管理其使用? 人工智能将如何影响就业和工作岗位?

要回答这些棘手的问题,技术专家需要与政府、学术界、企业、民间团体和其他利益相关者密切合作。在微软,我们确定了六个道德原则——公平性、可靠性和安全性、隐私和安全性、包容性、透明度、可说明性以指导跨学科开发和使用人工智能。当我们用社会规则管理人工智能时,我们越了解这些或类似的问题并且越多的技术开发人员和用户分享其解决这些问题的办法,我们越会得到更好的服务。

我们还必须关注人工智能对工人的影响。过去 250 年来不断发生的技术变化对就业产生了持续影响——新工作出现、现有工作消亡以及工作任务和内容变化。

一些关键结论正在出现。

首先,在人工智能时代,快速有效地接受这些变化的公司和国家会表现优异。这是因为那些拥抱技术的人会有新的就业机会和经济增长,而那些抵制或推迟采用该技术的人则不然。

其次,虽然我们认为人工智能将有助于解决重大的社会问题,但我们必须认识到,凡事都有两面性,既有挑战也有机遇。我们必须变革道德原则和法律,培训新技能甚至改革劳动力市场。如果我们要充分利用人工智能,就必须把这一切整合起来。

第三,我们需要有共同的责任感,因为人工智能不仅仅由科技部门创造。在微软,我们正在努力以类似于为每个人提供计算机的方式普及人工智能。这意味着我们正在创建工具,使每个开发人员、企业和政府都能轻松构建基于人工智能的解决方案并提升社会效益。

所有这些都会让我们得出一个最重要的结论。在人工智能驱动的世界中,技能提升不仅仅涉及科学、技术、工程和数学。随着计算机更像人类,社会科学和人文科学将变得更加重要。语言、艺术、历史、经济学、伦理学、哲学、心理学和人类发展课程可以培养基于批判、哲学和道德的技能,这些技能将有助于人工智能解决方案的开发和管理。如果人工智能要发挥为人类服务的潜力,那么每个工程师都需要更多地了解文科知识,每个文科生都需要更多地了解工程学。

新加坡培养未来就业人才的能力将成为数字经济蓬勃发展的关键,数字产品和服务将占经济的 60%。我们需要不断地重新评估并做好准备,以便迎接一个不断变化的未来世界。

然而,这个概念对于新加坡来说并不陌生,新加坡是一个多年来经历过并接受变革的年轻国家。新加坡的成功始终取决于其人民的技能、智慧和决心。在人工智能的帮助下,这种人类的聪明才智在未来几年将变得更加重要。

通过人工智能,我们现在所拥有的能力比我们前面的几代人都多。问题将是我们将如何利用它。

附录A

词汇表

A.1 单词表

单 词	音 标	意 义	课次
ability	[ə'bɪlɪtɪ]	n.能力,资格;能耐,才能	1a
absolute	['æbsəluːt]	adj.绝对的,完全的 n.绝对;绝对事物	6a
abstract	['æbstrækt]	adj.抽象的,理论上的	5a
abstraction	[æb'strækʃn]	n.抽象;抽象化;抽象概念	1b
academia	[ˌækə'diːmɪə]	n.学术界,学术环境	10a
accelerate	[ək'seləreɪt]	v.加快,加速	8a
acceleration	[əkˌselə'reɪʃn]	n.加速	7b
accelerator	[ək'seləreɪtə]	n.加速器	8a
accept	[ək'sept]	vi.承认,同意	5a
acceptance	[ək'septəns]	n.接受,接纳	9a
accident	['æksɪdənt]	n.意外事件;事故	1a
accommodate	[ə'kɒmədeɪt]	vt.容纳;使适应	6a
accountability	[əˌkaʊntə'bɪlɪtɪ]	n.有责任,有义务,责任制	10a
accuracy	['ækjərəsɪ]	n.精确(性),准确(性)	2b
accurate	['ækjʊrət]	adj.精确的,准确的;正确无误的	7a
acquisition	[ˌækwɪ'zɪʃn]	n.获得	1a
acquisition	[ˌækwɪ'zɪʃn]	n.获取,获得;收集	5a
action	['ækʃn]	n.行为,行动;功能,作用	2b
action	['ækʃn]	n.行动,活动;功能,作用;手段	5a
action	['ækʃn]	n.行动,活动;功能,作用	7a

续表

单 词	音 标	意 义	课次
adjust	[ə'dʒʌst]	v.调整,校正	2b
admissibility	[əd,mɪsə'bɪlətɪ]	n.可容许性,可接受性	2a
adopt	[ə'dɒpt]	vt.采用,采取	9a
adoption	[ə'dɒpʃn]	n.采用	1b
advancement	[əd'vɑːnsmənt]	n.前进,进步;提升,升级	9a
advice	[əd'vaɪs]	n.劝告,忠告;建议	6a
aerospace	['eərəuspeɪs]	n.航天	7b
affordable	[ə'fɔːdəbl]	adj.付得起的	10b
agent	['eɪdʒnt]	n.代理人;代理商 vt.由……作中介;由……代理 adj.代理的	1b
aid	[eɪd]	vt.辅助,帮助;促进 n.助手;辅助设备	5a
alarm	[ə'lɑːm]	n.警报 vt.警告	9a
algorithm	['ælɡərɪðəm]	n.算法	2a
alleviate	[ə'liːvɪeɪt]	vt.减轻,缓和	2b
alter	['ɔːltə]	v.改变	4a
altogether	[,ɔːltə'ɡeðə]	adv.全部地;完全地;总而言之	6b
ambience	['æmbɪəns]	n.气氛,周围环境	10a
amenity	[ə'miːnətɪ]	n.便利设施	9a
analogous	[ə'næləɡəs]	adj.类似的,相似的	4b
analyse	['ænəlaɪz]	vt.分析;分解	9a
analytic	[,ænə'lɪtɪk]	adj.分析的,解析的	1b
analyze	['ænəlaɪz]	vt.分析	8b
anatomy	[ə'nætəmɪ]	n.分解,分析	6a
animation	[,ænɪ'meɪʃn]	n.动画	4a
anneal	[ə'niːl]	n.退火 vt.使退火	2a
anomaly	[ə'nɒməli]	n.异常,反常	10b
anticipate	[æn'tɪsɪpeɪt]	vt.预感,预见,预料	10a
antithesis	[æn'tɪθɪsɪs]	n.对立,对立面	10b
app	[æp]	n.计算机应用程序 abbr.应用(Application)	9b
appointment	[ə'pɔɪntmənt]	n.预约	1a
apprehend	[,æprɪ'hend]	vt.逮捕,拘押	10b
approach	[ə'prəutʃ]	v.接近,走近,靠近 n.方法;途径;接近	5a
appropriate	[ə'prəuprɪət]	adj.适当的,恰当的,合适的	6a
approximate	[ə'prɒksɪmɪt]	adj.极相似的	1a
approximate	[ə'prɒksɪmeɪt]	vi.接近于,近似于 vt.靠近,接近	1a
approximately	[ə'prɒksɪmɪtlɪ]	adv.近似地,大约	7a
apt	[æpt]	adj.恰当的,适当的;聪明的,灵巧的	10b
arbitrary	['ɑːbɪtrərɪ]	adj.随意的,任性的	2a
arbitrary	['ɑːbɪtrərɪ]	adj.武断的,专断的	4a
arena	[ə'riːnə]	n.表演场地,舞台	9b
argument	['ɑːɡjumənt]	n.论据,理由,论点	6a

续表

单 词	音 标	意 义	课次
aspect	['æspekt]	n.方面；面貌	5a
assembler	[ə'semblə]	n.汇编程序	4a
assistance	[ə'sɪstəns]	n.帮助，援助；辅助设备	6a
assistant	[ə'sɪstənt]	n.助手，助理 adj.助理的；辅助的	9a
associate	[ə'səʊʃɪeɪt]	vt.(使)发生联系；(使)联合	6a
atmosphere	['ætməsfɪə]	n.大气，空气；大气层	2b
attacker	[ə'tækə]	n.攻击者	3b
attain	[ə'teɪn]	v.达到，实现；获得	7b
attendee	[ˌæten'diː]	n.(会议等的)出席者	9a
attention	[ə'tenʃn]	n.注意，注意力	1b
attract	[ə'trækt]	vt.吸引；引起…的好感(或兴趣)vi.具有吸引力；引人注意	7a
aural	['ɔːrəl]	adj.耳的；听觉的	5a
authenticate	[ɔː'θentɪkeɪt]	vt.认证；鉴定	3b
authentication	[ɔːˌθentɪ'keɪʃn]	n.身份验证；认证	9a
authenticity	[ˌɔːθen'tɪsɪtɪ]	n.可靠性，确实性，真实性	3b
authorize	['ɔːθəraɪz]	vt.授权，批准	3b
automatically	[ˌɔːtə'mætɪklɪ]	adv.自动地	7b
automation	[ˌɔːtə'meɪʃn]	n.自动化(技术)，自动操作	1b
automobile	['ɔːtəməbiːl]	n.汽车	9a
automotive	[ˌɔːtə'məʊtɪv]	adj.汽车的；自动的	7b
autonomous	[ɔː'tɒnəməs]	adj.自治的,有自主权的	
avatar	['ævətɑː]	n.化身	2b
backend	[bækend]	n.后端	7a
back-propagation	[bæk-prɒpə'geɪʃn]	n.反向传播	8a
backslash	['bækslæʃ]	n.反斜线符号	4b
beforehand	[bɪ'fɔːhænd]	adv.事先,预先；提前 adj.提前的；预先准备好的	10b
behave	[bɪ'heɪv]	vi.表现	3b
behavior	[bɪ'heɪvjə]	n.行为；态度	2b
benefit	['benɪfɪt]	n.利益,好处 vt.有益于,有助于；使受益；得益	2b
betterment	['betəmənt]	n.改良,改进	5a
bias	['baɪəs]	n.偏见；倾向	6a
bidirectional	[ˌbaɪdɪ'rekʃənl]	adj.双向的	2a
biologically	[ˌbaɪəʊ'lɒdʒɪkəlɪ]	adv.生物学上地	8a
biometric	[ˌbaɪəʊ'metrɪk]	n.计量生物学	1b
biometrics	[ˌbaɪəʊ'metrɪks]	n.生物测定学	3b
blast	[blɑːst]	vi.严厉批评或猛烈攻击	10b
blur	[blɜː]	v.模糊	4a
bootstrap	['buːtstræp]	n.引导	3b

续表

单　词	音　标	意　义	课次
bored	[bɔːd]	adj.无聊的,无趣的,烦人的	2b
bottleneck	['bɒtlnek]	n.瓶颈	5b
boundary	['baʊndrɪ]	n.边界,分界线;范围	5a
breakdown	['breɪkdaʊn]	n.损坏,故障;崩溃,倒塌	2b
breakthrough	['breɪkθruː]	n.突破;重要技术成就	10a
budget	['bʌdʒɪt]	n.预算;预算拨款 v.把……编入预算	1b
bulk	[bʌlk]	n.大块,大量;大多数,大部分	7a
bundle	['bʌndl]	n.捆,束,包 v.捆扎	4a
calendar	['kælɪndə]	n.日程表	9a
call	[kɔːl]	n.&v.调用	4a
camera	['kæmərə]	n.摄像头,摄影机;照相机	10b
cancel	['kænsl]	vt.取消,注销	9a
candidate	['kændɪdət]	n.候选者,候选人	3a
capable	['keɪpəbl]	adj.有能力的;熟练的;胜任的	5a
capture	['kæptʃə]	vt.&n.捕获,捕捉	6a
catalyst	['kætəlɪst]	n.催化剂,促进因素	10a
categorize	['kætəgəraɪz]	vt.把…归类,把…分门别类	5b
cell	[sel]	n.细胞	7b
centroid	['sentrɔɪd]	n.质心;矩心	8b
chance	[tʃɑːns]	n.机会,机遇	6a
chaos	['keɪɒs]	n.混乱,紊乱;一团糟	2b
character	['kærɪktə]	n.角色,人物;性格;特点;字母	3a
chatbot	[tʃætbɒt]	n.聊天机器人	1a
check	[tʃek]	vt.检查,核对	2a
chip	[tʃɪp]	n.芯片	9a
chum	[tʃʌm]	vi.结交,成为好朋友 n.密友	6a
circumstance	['sɜːkəmstəns]	n.环境,境遇	1a
class	[klɑːs]	n.类	4b
classification	[ˌklæsɪfɪ'keɪʃn]	n.分类;分级;类别	1b
clause	[klɔːz]	n.子句	4b
client	['klaɪənt]	n.顾客;当事人;[计算机]客户端	9a
clipboard	['klɪpbɔːd]	n.剪贴板	4a
clustering	['klʌstərɪŋ]	n.聚类	8b
co-bot	[ˌkəʊ-bɒt]	n.协作机器人	9b
coefficient	[ˌkəʊɪ'fɪʃnt]	n.系数	3b
cognition	[kɒg'nɪʃn]	n.认识,认知	5a
cognitive	['kɒgnɪtɪv]	adj.认知的,认识的	1a
collaboration	[kəˌlæbə'reɪʃn]	n.合作,协作	5b
collaborator	[kə'læbəreɪtə]	n.协作者,合作者	6b
collapse	[kə'læps]	vi.(尤指工作劳累后)坐下	10b
collection	[kə'lekʃn]	n.收集,采集	3b

单　词	音　标	意　义	课次
combination	[ˌkɒmbɪˈneɪʃn]	n.结合；联合体	6a
combine	[kəmˈbaɪn]	v.组合，使结合	5b
comfortable	[ˈkʌmftəbl]	adj.舒适的	9a
comma	[ˈkɒmə]	n.逗号	7a
commitment	[kəˈmɪtmənt]	n.承诺，许诺；委任，委托	1a
commodity	[kəˈmɒdəti]	n.商品，日用品	10a
communication	[kəˌmjuːnɪˈkeɪʃn]	n.通信	4b
community	[kəˈmjuːnɪti]	n.社区，社团	5b
competently	[ˈkɒmpɪtəntlɪ]	adv.胜任地，适合地	4a
competitor	[kəmˈpetɪtə]	n.竞争者；对手	3b
compile	[kəmˈpaɪl]	vt.编译	4a
complexity	[kəmˈpleksɪti]	n.复杂度，复杂性	2a
complication	[ˌkɒmplɪˈkeɪʃn]	n.纠纷；混乱	6b
composition	[ˌkɒmpəˈzɪʃn]	n.组合方式；成分；构成	2b
comprehensive	[ˌkɒmprɪˈhensɪv]	adj.广泛的；综合的	7a
compromise	[ˈkɒmprəmaɪz]	n.损害；妥协，折中方案 vi.折中解决；妥协	3b
conceive	[kənˈsiːv]	v.构思；想象，设想	6a
concept	[ˈkɒnsept]	n.观念，概念；观点；思想	5a
conceptualization	[kənˈseptjʊəlaɪˈzeɪʃən]	n.化为概念，概念化	6a
conclude	[kənˈkluːd]	vt.得出结论；推断出；决定	7a
conclusion	[kənˈkluːʒn]	n.结论，断定，决定；推论	1a
conclusion	[kənˈkluːʒn]	n.断定，推论；结论，结局	6b
conditionally	[kənˈdɪʃənəlɪ]	adv.有条件地	4a
conduct	[kənˈdʌkt]	v.引导；实施；执行	6a
configure	[kənˈfɪgə]	v.配置；设定	3b
confusion	[kənˈfjuːʒn]	n.混乱，混淆；困惑	5b
congestion	[kənˈdʒestʃən]	n.拥挤，堵车	10b
connotation	[ˌkɒnəˈteɪʃn]	n.内涵，含义	1a
consciousness	[ˈkɒnʃəsnɪs]	n.意识，观念；知觉	1a
considerable	[kənˈsɪdərəbl]	adj.相当大（或多）的；该注意的，应考虑的	8a
considerably	[kənˈsɪdərəblɪ]	adv.相当，非常，颇	6a
consideration	[kənˌsɪdəˈreɪʃn]	n.考虑，考察，照顾，关心	6a
consistency	[kənˈsɪstənsɪ]	n.一致性；符合；前后一致	5a
consistent	[kənˈsɪstənt]	adj.一致的；连续的；坚持的	5b
consistently	[kənˈsɪstəntlɪ]	adv.一贯地，坚持地	1a
consume	[kənˈsjuːm]	vt.消耗，消费	2a
consumption	[kənˈsʌmpʃn]	n.消费	10a
contact	[ˈkɒntækt]	n.接触 vt.使接触；与……联系	5b
contemplate	[ˈkɒntəmpleɪt]	vt.周密考虑 vi.沉思，深思熟虑	10a
context	[ˈkɒntekst]	n.上下文；背景；语境	9a
continuous	[kənˈtɪnjʊəs]	adj.连续的；延伸的；不断的	8b

续表

单　词	音　标	意　义	课次
continuously	[kən'tɪnjuəslɪ]	adv.连续不断地,接连地	3b
contraction	[kən'trækʃn]	n.缩写式,紧缩	4b
contributor	[kən'trɪbjʊtə]	n.贡献者;捐助者;投稿者	5b
convenient	[kən'viːnɪənt]	adj.便利的,方便的	4a
converge	[kən'vɜːdʒ]	vi.收敛	7b
conversation	[ˌkɒnvə'seɪʃn]	n.交谈,会话;(人与计算机的)人机对话	9a
convert	[kən'vɜːt]	v.转变	9a
convincingly	[kən'vɪnsɪŋlɪ]	adv.令人信服地,有说服力地	1a
convolutional	[ˌkɒnvə'luːʃənl]	adj.卷积的	8a
convolve	[kən'vɒlv]	v.卷积	7b
coordination	[kəʊˌɔːdɪ'neɪʃn]	n.协调	9a
coroutine	[kəruː'tiːn]	n.协同程序	4b
correction	[kə'rekʃn]	n.修改 adj.改正的,纠正的	6b
correlation	[ˌkɒrə'leɪʃn]	n.相互关系;相关性	9a
correspondence	[ˌkɒrə'spɒndəns]	n.一致,符合;对应	5a
costly	['kɒstlɪ]	adj.昂贵的,困难的;造成损失的	4a
countless	['kaʊntlɪs]	adj.无数的,多得数不清的	6a
co-worker	['kəʊwɜːkə]	n.合作者,同事,帮手	4a
crawler	['krɔːlə]	n.爬虫,爬行动物	3a
creation	[krɪ'eɪʃn]	n.制造,创造	5b
creativity	[ˌkriːeɪ'tɪvɪtɪ]	n.创造性,创造力	10b
credit	['kredɪt]	n.信誉,信用;[金融]贷款 vt.相信,信任	1a
critical	['krɪtɪkl]	adj.关键的;极重要的	3b
cross-disciplinary	[krɒs-'dɪsəplɪnərɪ]	n.跨学科	10a
crossword	['krɒswɜːd]	n.填字游戏,纵横字谜	2a
crucial	['kruːʃl]	adj.关键性的,极其显要的;决定性的	7b
crunch	[krʌntʃ]	vt.快速处理	9a
cryptographic	['krɪptəʊ'græfɪk]	adj.加密的,用密码写的	3b
curate	['kjʊəreɪt]	v.管理	5b
cut-off	['kʌtɔf]	n.截止;界限	2a
damage	['dæmɪdʒ]	n.损害,损毁;赔偿金 v.损害,毁坏	1a
dangerous	['deɪndʒərəs]	adj.危险的	2b
database	['deɪtəbeɪs]	n.数据库	5a
debugger	[ˌdiː'bʌgə]	n.调试器	4a
debugging	['diːbʌgɪŋ]	n.调试	4b
decide	[dɪ'saɪd]	vt.决定;解决;裁决 vi.决定;下决心	7a
decision	[dɪ'sɪʒn]	n.决定,决策	1b
decision-making	[dɪ'sɪʒn'meɪkɪŋ]	n.决策 adj.决策的	1b
declared	[dɪ'kleəd]	adj.声明的	4b
decline	[dɪ'klaɪn]	n.衰退;下降	2b
dedicated	['dedɪkeɪtɪd]	adj.专注的,专用的	10b

续表

单　词	音　标	意　义	课次
dedication	[ˌdedɪˈkeɪʃn]	n.奉献,献身精神	2b
deductive	[dɪˈdʌktɪv]	adj.推论的,演绎的	5b
deepfake	[ˈdiːpfeɪk]	n.换脸术	1a
defense	[dɪˈfens]	n.国防,防卫	7b
definite	[ˈdefɪnɪt]	adj.明确的；一定的；肯定	1a
definition	[ˌdefɪˈnɪʃn]	n.定义；解释	3a
definitively	[dɪˈfɪnɪtɪvlɪ]	adv.决定性地,最后地	6a
delete	[dɪˈliːt]	v.删除	7a
deliver	[dɪˈlɪvə]	vt.交付,递送	5b
democratise	[dɪˈmɒkrətaɪz]	v.民主化	10a
demonstrate	[ˈdemənstreɪt]	vt.证明,证实；显示,展示	10b
deploy	[dɪˈplɔɪ]	v.使展开；施展；有效地利用	7a
depression	[dɪˈpreʃn]	n.萎靡不振,沮丧	2b
depth	[depθ]	n.深度	2a
description	[dɪˈskrɪpʃn]	n.描述,形容	2a
design	[dɪˈzaɪn]	v.&n.设计	1b
desirable	[dɪˈzaɪərəbl]	adj.令人满意的；值得拥有的；可取的	8a
destination	[ˌdestɪˈneɪʃn]	n.目的,目标	3a
destruction	[dɪˈstrʌkʃn]	n.破坏,毁灭,消灭,灭亡	2b
destructive	[dɪˈstrʌktɪv]	adj.破坏性的；毁灭性的；有害的	6a
detail	[ˈdiːteɪl]	n.细节；详述	5a
detect	[dɪˈtekt]	vt.检测,发现	7b
detection	[dɪˈtekʃn]	n.检查,检测	1a
determination	[dɪˌtɜːmɪˈneɪʃn]	n.决定,确定	3a
determine	[dɪˈtɜːmɪn]	vt.决定,确定	3a
developer	[dɪˈveləpə]	n.开发者	10a
development	[dɪˈveləpmənt]	n.发展,进化	6a
diagnose	[ˈdaɪəgnəuz]	vt.诊断；判断 vi.做出诊断	1a
diagnosis	[ˌdaɪəgˈnəusɪs]	n.诊断；判断；结论	2b
diagnostic	[ˌdaɪəgˈnɒstɪk]	adj.诊断的,判断的；特征的	6a
difficulty	[ˈdɪfɪkəltɪ]	n.难度；困难,麻烦	6a
dig	[dɪg]	vt.挖掘；发掘	2b
digitized	[ˈdɪdʒɪtaɪzd]	v.数字化	6a
diminish	[dɪˈmɪnɪʃ]	vt.(使)减少,缩小 vi.变小或减少	2b
direction	[dəˈrekʃn]	n.方向	8a
disadvantage	[ˌdɪsədˈvɑːntɪdʒ]	n.缺点,劣势,短处	2a
disclosure	[dɪsˈkləuʒə]	n.泄露,揭露	3b
discrete	[dɪˈskriːt]	adj.分离的,不相关联的	6a
discuss	[dɪˈskʌs]	vt.讨论,谈论；论述,详述	3a
disk	[dɪsk]	n.磁盘	3b
disposal	[dɪˈspəuzl]	n.(事情的)处置；清理 adj.处理废品的	3b

续表

单 词	音 标	意 义	课次
dispose	[dɪˈspəʊz]	v.处理,处置;安排	2a
disseminate	[dɪˈsemɪneɪt]	vt.散布,传播	9b
dissonance	[ˈdɪsənəns]	n.不一致,不和谐	5a
distinguish	[dɪˈstɪŋgwɪʃ]	vi.区分,辨别,分清	6b
distribute	[dɪˈstrɪbjuːt]	vt.分配,散布;散发,分发	6a
disturbance	[dɪˈstɜːbəns]	n.打扰,困扰	6a
documentation	[ˌdɒkjʊmenˈteɪʃn]	n.文档	5a
download	[ˌdaʊnˈləʊd]	v.下载	3b
downtime	[ˈdaʊntaɪm]	n.停工期	6a
dramatic	[drəˈmætɪk]	adj.戏剧性的;引人注目的;突然的;巨大的	9a
drone	[drəʊn]	n.无人机	6b
drop-off	[ˈdrɒpɒf]	n.急下降,直下降	10b
duplicate	[ˈdjuːplɪkeɪt]	v.重复	2a
easy-to-use	[ˈiːzɪ-tʊ-juːz]	adj.易用的,好用的	5b
edge	[edʒ]	n.边	2a
efficiency	[ɪˌfekˈtɪvnɪs]	n.有效性;有效,有力	3a
elaborate	[ɪˈlæbərət]	vi.详尽说明;变得复杂 vt.详细制定;详尽阐述	9b
email	[ˈiːmeɪl]	n.电子邮件 vt.给……发电子邮件	9a
embrace	[ɪmˈbreɪs]	v.拥抱;包括,包含;接受	9b
emerge	[ɪˈmɜːdʒ]	vi.出现,浮现,涌现	6a
emotion	[ɪˈməʊʃn]	n.情感,情绪	2b
emotional	[ɪˈməʊʃənl]	adj.表现强烈情感的;令人动情的;易动感情的	2b
emphasize	[ˈemfəsaɪz]	vt.强调,着重;使突出	6a
empower	[ɪmˈpaʊə]	vt.授权,准许;使能够	9b
empty	[ˈemptɪ]	adj.空的 vt.(使)成为空的	3a
enablement	[ɪˈneɪblmənt]	n.允许,启动,实现	3b
enclose	[ɪnˈkləʊz]	vt.封装	4b
encode	[ɪnˈkəʊd]	vt.译成密码;编码	5a
encompass	[ɪnˈkʌmpəs]	vt.围绕,包围;包含	6a
encounter	[ɪnˈkaʊntə]	vt.遭遇;对抗 n.对决,冲突;相遇,碰见	5a
encourage	[ɪnˈkʌrɪdʒ]	vt.鼓励,支持	10b
encyclopedia	[ɪnˌsaɪkləˈpiːdɪə]	n.百科全书	5b
endeavor	[ɪnˈdevə]	n.&vi.尽力,努力	4a
end-to-end	[end-tuː-end]	adj.端到端的,端对端的	3b
endure	[ɪnˈdjʊə]	vt.忍耐;容忍	2b
enemy	[ˈenəmɪ]	n.敌军 adj.敌人的;敌方的	3a
enforce	[ɪnˈfɔːs]	vt.实施,执行;加强	3b
engine	[ˈendʒɪn]	n.引擎,发动机	6a

续表

单　词	音　标	意　义	课次
enhance	[ɪn'hɑːns]	vt.提高,增加;加强	6a
enormous	[ɪ'nɔːməs]	adj.巨大的,庞大的	2b
enormously	[ɪ'nɔːməslɪ]	adv.巨大地,庞大地	6a
enslave	[ɪn'sleɪv]	vt.奴役;征服	2b
ensure	[ɪn'ʃʊə]	vt.确保	3b
environment	[ɪn'vaɪrənmənt]	n.环境,外界	10b
essential	[ɪ'senʃl]	adj.基本的;必要的;本质的 n.必需品;基本要素	3b
establish	[ɪ'stæblɪʃ]	vt.建立,创建	3b
estimate	['estɪmɪt]	n.估计,预测	1b
estimate	['estɪmeɪt]	vt.估计,估算;评价	2a
ethic	['eθɪk]	n.伦理,道德	10a
ethical	['eθɪkl]	adj.道德的,伦理的	1a
euclidean	[juː'klɪdɪən]	adj.欧几里得的,欧几里得几何学的	8b
evaluation	[ɪ,væljʊ'eɪʃn]	n.评估,估价	2a
evidence	['evɪdəns]	n.证据;迹象 vt.显示;表明;证实	6a
evolution	[,iːvə'luːʃn]	n.演变,进化,发展	10a
evolve	[i'vɒlv]	vt.使发展;使进化;设计	5a
exacerbate	[ɪg'zæsəbeɪt]	vt.使恶化;使加重	10b
exactly	[ɪg'zæktlɪ]	adv.精确地;确切地	8a
examination	[ɪg,zæmɪ'neɪʃn]	n.检查	6a
exceed	[ɪk'siːd]	vt.超过,超越,胜过	7b
exception	[ɪk'sepʃn]	n.异常,例外	4b
excite	[ɪk'saɪt]	vt.使兴奋;激发;刺激;使紧张不安	2b
execution	[,eksɪ'kjuːʃn]	n.执行,完成	9a
exercise	['eksəsaɪz]	vi.训练,练习 n.练习,训练;运用	6a
exit	['eksɪt]	n.出口,退出 vi.离开;退出	3a
expectation	[,ekspek'teɪʃn]	n.期待;预期	1a
expensive	[ɪk'spensɪv]	adj.昂贵的,花钱多的	6a
experience	[ɪk'spɪərɪəns]	n.体验,经验;经历,阅历	2b
experiential	[ɪk,spɪərɪ'enʃl]	adj.经验的,经验上的,根据经验的	6a
experiment	[ɪk'sperɪmənt]	n.实验,试验;尝试 vi.做实验	1a
expertise	[,ekspɜː'tiːz]	n.专门知识或技能;专家的意见;专家评价	5a
explain	[ɪk'spleɪn]	v.说明,解释	
explanation	[,eksplə'neɪʃn]	n.解释;说明	6a
explicitly	[ɪk'splɪsɪtlɪ]	adv.明白地,明确地	7b
exponential	[,ekspə'nenʃl]	adj.指数的,幂数的;越来越快的 n.指数	2a
exponentiation	[,ekspəʊ,nenʃɪ'eɪʃən]	n.求幂	4b
expression	[ɪk'spreʃn]	n.表达式	4b
extent	[ɪk'stent]	n.程度;长度	3a
extract	['ɪkstrækt]	vt.提取;选取;获得	6a

续表

单　词	音　标	意　义	课次
extraction	[ɪk'strækʃn]	n.取出，提取	7b
extremely	[ɪk'striːmlɪ]	adv.极端地；非常，很	6a
fabricate	['fæbrɪkeɪt]	vt.编造，捏造	1a
facial	['feɪʃl]	adj.面部的；表面的	3b
facilitate	[fə'sɪlɪteɪt]	vt.促进，助长	1b
factor	['fæktə]	n.因素，因子	2a
factual	['fæktʃʊəl]	adj.事实的，真实的	6a
failure	['feɪljə]	n.失败	3a
fairness	['feənɪs]	n.公正，公平	10a
fantastic	[fæn'tæstɪk]	adj.极好的；很大的	9b
fear	[fɪə]	n.害怕；可能性 vt.害怕；为……忧虑(或担心、焦虑)vi.害怕；忧虑	1a
feed	[fiːd]	vt.馈送；向……提供	7a
feedforward	['fiːdfɔːwəd]	n.前馈	8a
filter	['fɪltə]	n.过滤器，滤波器	4a
fine-tune	[faɪn-tjuːn]	vt.微调，调整	7b
fingerprint	['fɪŋgəprɪnt]	n.指纹，指印 vt.采指纹	3b
fingertip	['fɪŋgətɪp]	n.指尖	10a
firm	[fɜːm]	n.公司，企业 adj.坚固的，确定的	10a
firmware	['fɜːmweə]	n.(计算机的)固件	3b
flourish	['flʌrɪʃ]	vi.茂盛，繁荣；活跃，蓬勃	1b
folk	[fəʊk]	n.人们；各位；大伙儿	5b
font	[fɒnt]	n.字体	5b
forefront	['fɔːfrʌnt]	n.前列；第一线；活动中心	1a
formal	['fɔːml]	adj.规则的，正规的；形式的	6a
formalization	[ˌfɔːməlaɪ'zeɪʃn]	n.形式化；规则化	6a
format	['fɔːmæt]	n.格式 vt.使格式化	1b
foundational	[faʊn'deɪʃənl]	adj.基本的，基础的	3b
fragmented	[fræg'mentɪd]	adj.成碎片的，片断的	5a
frame	[freɪm]	n.框架	6a
framework	['freɪmwɜːk]	n.构架；框架	3b
framing	['freɪmɪŋ]	n.构架，框架，骨架	9b
fraud	[frɔːd]	n.欺诈；骗子；伪劣品；冒牌货	8a
frequency	['friːkwənsɪ]	n.频率，次数；频率分布	2a
fuel	['fjuːəl]	n.燃料	2b
function	['fʌŋkʃn]	n.函数	7a
functionality	[ˌfʌŋkʃə'nælɪtɪ]	n.功能性	3b
fundamentally	[ˌfʌndə'mentəlɪ]	adv.基础地；根本地；从根本上	6b
fuse	[fjuːz]	vi.融化；融合 vt.使融合；使融化	6a
fusion	['fjuːʒn]	n.融合	9b
futuristic	[ˌfjuːtʃə'rɪstɪk]	adj.未来的；未来派的；未来主义的	9b

续表

单　词	音　标	意　义	课次
fuzzy	[ˈfʌzɪ]	adj.模糊的	6a
gain	[geɪn]	v.获得；赢得	5a
game-changer	[geɪm-ˈtʃeɪndʒə]	n.规则改变者；打破格局、扭转局面的事物	9a
gatekeeper	[ˈgeɪtkiːpə]	n.看门人	5b
gateway	[ˈgeɪtweɪ]	n.门；入口；途径	9a
gauge	[geɪdʒ]	n.评估 vt.评估，判断	9a
general-purpose	[ˈdʒenrəl-ˈpɜːpəs]	adj.多种用途的	4b
generate	[ˈdʒenəreɪt]	vt.产生，造成；形成	5a
generation	[ˌdʒenəˈreɪʃn]	n.产生，生成	1b
generator	[ˈdʒenəreɪtə]	n.生成器	4b
generic	[dʒəˈnerɪk]	adj.类的，属性的；一般的	9b
gesture	[ˈdʒestʃə]	n.手势 vt.做手势	8b
governance	[ˈgʌvənəns]	n.管理；支配	1b
gradually	[ˈgrædʒuəlɪ]	adv.逐步地，渐渐地	7a
graph	[grɑːf]	n.图表，曲线图 vt.用曲线图表示	2a
guarantee	[ˌgærənˈtiː]	n.保证，担保；保证人，保证书 vt.保证，担保	3a
guidance	[ˈgaɪdns]	n.指导，引导	5b
guideline	[ˈgaɪdlaɪn]	n.指导方针；指导原则	3b
hacker	[ˈhækə]	n.(电脑)黑客	3b
hallmark	[ˈhɔːlmɑːk]	n.特点，标志 vt.使具有…标志	9a
handle	[ˈhændl]	v.操作，操控 n.手柄；句柄	5a
handwriting	[ˈhændraɪtɪŋ]	n.书法，手书；笔迹，字迹	8a
hardware	[ˈhɑːdweə]	n.计算机硬件	3b
harness	[ˈhɑːnɪs]	vt.利用；控制	2b
harvest	[ˈhɑːvɪst]	n.收成；结果 v.收割，收成	10a
headline	[ˈhedlaɪn]	n.大字标题；新闻提要；头条新闻	10a
healthcare	[ˈhelθkeə]	n.卫生保健	1a
heuristic	[hjʊˈrɪstɪk]	adj.启发式的；探试的，探索的	2a
high-level	[ˈhaɪ-ˈlevəl]	adj.高级的	4b
high-performance	[haɪ-pəˈfɔːməns]	adj.高性能的	7b
high-throughput	[haɪ-ˈθruːpʊt]	adj.高吞吐量的	8a
hindrance	[ˈhɪndrəns]	n.妨害，障碍；障碍物	2b
horizontally	[ˌhɒrɪˈzɒntəlɪ]	adv.水平地，横地	7a
humanity	[hjuːˈmænətɪ]	n.人文学科	10a
human-readable	[ˈhjuːmən-ˈriːdəbl]	adj.人可读的	5b
hype	[haɪp]	n.天花乱坠的广告宣传 vt.大肆宣传；夸张地宣传	1b
hyperlink	[ˈhaɪpəlɪŋk]	n.超链接	3a
hypothesis	[haɪˈpɒθəsɪs]	n.假设，假说；前提	1a
identical	[aɪˈdentɪkl]	adj.同一的，相同的	8a
identifiable	[aɪˌdentɪˈfaɪəbl]	adj.可辨认的，可识别的	3b

单 词	音 标	意 义	课次
identification	[aɪˌdentɪfɪˈkeɪʃn]	n.鉴定,识别	1a
ignorance	[ˈɪgnərəns]	n.无知,愚昧	6a
imagination	[ɪˌmædʒɪˈneɪʃn]	n.想象,想象力	9a
imagine	[ɪˈmædʒɪn]	vt.设想;想象;猜想;误认为 vi.想象;猜想,推测	10b
imitate	[ˈɪmɪteɪt]	vt.模仿,效仿	6a
immediate	[ɪˈmiːdɪət]	adj.最接近的;立即的;直接的	3a
immense	[ɪˈmens]	adj.极大的,巨大的	9b
immutable	[ɪˈmjuːtəbl]	adj.不可变的,不能变的	4b
impede	[ɪmˈpiːd]	vt.阻碍;妨碍;阻止	1a
imperative	[ɪmˈperətɪv]	n.命令 adj.命令的	4b
implement	[ˈɪmplɪment]	vt.实施,执行;使生效,实现 n.工具,手段	1b
implementation	[ˌɪmplɪmenˈteɪʃn]	n.执行,履行;落实	6a
implication	[ˌɪmplɪˈkeɪʃn]	n.含义,含意	9b
impose	[ɪmˈpəʊz]	vi.利用;施加影响	8a
impressive	[ɪmˈpresɪv]	adj.给人印象深刻的,引人注目的;可观的	10a
improper	[ɪmˈprɒpə]	adj.不适当的,不合适的,不正确的	4b
improvement	[ɪmˈpruːvmənt]	n.改进,改善,改良,增进	2a
inaccurate	[ɪnˈækjʊrɪt]	adj.不精确的;不准确,不正确的	6a
inadequate	[ɪnˈædɪkwɪt]	adj.不适当的;不足胜任的;不充足的	6a
incident	[ˈɪnsɪdənt]	n.事件	6a
include	[ɪnˈkluːd]	vt.包括,包含	3a
inclusivity	[ˌɪnkluːˈsɪvɪtɪ]	n.包容性	10a
incorporate	[ɪnˈkɔːpəreɪt]	v.包含;合并;混合	5a
incorporating	[ɪnˈkɔːpəreɪtɪŋ]	v.融合,包含;使混合	1a
incrementally	[ˌɪnkrɪˈmentəlɪ]	adv.逐渐地	2a
incur	[ɪnˈkɜː]	vt.引起,招致	2b
indentation	[ˌɪndenˈteɪʃn]	n.缩排	4b
indicate	[ˈɪndɪkeɪt]	vt.表明,标示,指示	5a
individual	[ˌɪndɪˈvɪdʒuəl]	adj.个人的;独特的;个别的 n.个人;个体	2b
inevitably	[ɪnˈevɪtəblɪ]	adv.必然地,无疑的	10a
infer	[ɪnˈfɜː]	vt.推断;猜想,推理 vi.作出推论	3b
infinitely	[ˈɪnfɪnətlɪ]	adv.无限地,无穷地	2a
infix	[ˈɪnfɪks]	n.中缀 vt.让……插进	4b
inform	[ɪnˈfɔːm]	vt.通知	2a
infuse	[ɪnˈfjuːz]	vt.灌输,使充满;鼓舞,激发	9b
inherent	[ɪnˈhɪərənt]	adj.固有的,内在的;天生	2b
inheritance	[ɪnˈherɪtəns]	n.继承;遗传	6a
initialization	[ɪˌnɪʃəlaɪˈzeɪʃn]	n.设定初值,初始化	8b
initiative	[ɪˈnɪʃətɪv]	n.主动性;主动权 adj.自发的;创始的	10b
innovation	[ˌɪnəˈveɪʃn]	n.改革,创新;新观念,新发明	3b

单　词	音　标	意　义	课次
input	['ɪnpʊt]	n.输入	5a
insight	['ɪnsaɪt]	n.洞察力,洞悉	1b
inspire	[ɪn'spaɪə]	vt.赋予灵感,启发,启迪;激励	8a
instance	['ɪnstəns]	n.例子,实例;情况;要求,建议	8b
institute	['ɪnstɪtjuːt]	vt.建立;制定	5b
instructive	[ɪn'strʌktɪv]	adj.有益的;教育性的 adv.有益地;有教益地 n.教育;指导性	6a
instrumental	[ˌɪnstrə'mentl]	adj.有帮助的,起作用的	10a
insulate	['ɪnsjʊleɪt]	vt.隔离,使绝缘	4a
intact	[ɪn'tækt]	adj.完整无缺的,未受损伤的;原封不动的;	3b
integrative	['ɪntɪgreɪtɪv]	adj.综合的,一体化的	1a
integrity	[ɪn'tegrɪtɪ]	n.完整,完整性	3b
intellectual	[ˌɪntə'lektʃuəl]	adj.智力的,有才智的 n.知识分子;脑力劳动者	2b
intended	[ɪn'tendɪd]	adj.有意的,预期的	6a
intention	[ɪn'tenʃn]	n.意图,目的;意向	1a
interaction	[ˌɪntər'ækʃn]	n.合作;互相影响;互动	6a
interactive	[ˌɪntər'æktɪv]	adj.交互式的,互动的;互相作用的,相互影响的	1b
intercept	[ˌɪntə'sept]	vt.拦截,拦阻	2b
interconnect	[ˌɪntəkə'nekt]	vi.互相连接,互相联系 vt.使互相连接;使互相联系	1b
interconnection	[ˌɪntəkə'nekʃn]	n.互相连接	4a
interference	[ˌɪntə'fɪərəns]	n.干涉,干扰,冲突;妨碍	9b
intermediate	[ˌɪntə'miːdɪət]	adj.中间的,中级的 n.中间物,中间人 vi.调解;干涉	8a
internal	[ɪn'tɜːnl]	adj.内部的	2a
interpretation	[ɪnˌtɜːprɪ'teɪʃn]	n.解释,说明	6a
interpreter	[ɪn'tɜːprɪtə]	n.解释程序	4a
interpretive	[ɪn'tɜːprɪtɪv]	adj.作为说明的,解释的	6a
interrupt	[ˌɪntə'rʌpt]	v.&n.中断;暂停	2a
intervention	[ˌɪntə'venʃn]	n.介入,干涉,干预	7a
intrinsic	[ɪn'trɪnsɪk]	adj.固有的,内在的,本质的	6a
introduction	[ˌɪntrə'dʌkʃn]	n.介绍;引言,导言	3b
invaluable	[ɪn'væljʊəbl]	adj.非常宝贵的;无法估计的;无价的	5b
inverse	[ˌɪn'vɜːs]	adj.相反的;逆向的;倒转的 n.相反;倒转;相反的事物 vt.使倒转	2a
investment	[ɪn'vestmənt]	n.投资,投资额	1b
involve	[ɪn'vɒlv]	vt.包含;使参与,牵涉	5a
iterate	['ɪtəreɪt]	vt.重复	4b
iterative	['ɪtərətɪv]	adj.迭代的,重复的,反复的 n.反复体	2a

续表

单　词	音　标	意　义	课次
iteratively	['ɪtə‚reɪtɪvlɪ]	adv.反复地；迭代地	4a
jobless	['dʒɒbləs]	adj.没有工作的，失业的	2b
judge	[dʒʌdʒ]	v.评判；断定	2b
junk	[dʒʌŋk]	vt.丢弃，废弃 n.废品；假货	1a
kangaroo	[‚kæŋgə'ruː]	n.袋鼠	5a
keyword	['kiːwɜːd]	n.关键字	4b
knowledge	['nɒlɪdʒ]	n.知识；了解，理解	7a
laborious	[lə'bɔːrɪəs]	adj.费力的；勤劳的；辛苦的	2b
lamppost	['læmppəʊst]	n.灯杆，路灯柱	7b
large-scale	[lɑːdʒ-skeɪl]	adj.大规模的，大范围的	5a
latency	['leɪtənsɪ]	n.延迟；潜伏	8a
lawmaker	['lɔːmeɪkə]	n.立法者	1a
layer	['leɪə]	n.层，层次 vt.把……分层	1b
legislation	[‚ledʒɪs'leɪʃn]	n.立法；法律，法规	10b
lengthy	['leŋθɪ]	adj.长的，漫长的	6a
lifecycle	['laɪf‚saɪkl]	n.生命周期	3b
likelihood	['laɪklɪhʊd]	n.可能，可能性；[数]似然，似真	6a
linear	['lɪnɪə]	adj.线性的	2a
link	[lɪŋk]	vt.链接，连接 n.链接；关联	6a
linker	['lɪŋkə]	n.(目标代码)连接器	4a
list	[lɪst]	n.列表；清单，目录 vt.列出	2a
literal	['lɪtərəl]	adj.文字的，照字面上的	4b
log	[lɒg]	n.记录；日志	9a
low-latency	[ləʊ-'leɪtənsɪ]	adj.低延迟的	8a
machine-readable	[mə'ʃiːn-'riːdəbl]	adj.机器可读的	5b
machinery	[mə'ʃiːnərɪ]	n.机器，机械装置	6b
macro	['mækrəʊ]	n.宏	4a
manageable	['mænɪdʒəbl]	adj.可管理的，易处理的，易控制的	9a
manager	['mænɪdʒə]	n.管理器	4b
manual	['mænjʊəl]	n.手册；指南 adj.用手的；手制的，手工的	5b
manufacturer	[‚mænjʊ'fæktʃərə]	n.制造商，制造厂	3b
map	[mæp]	vt.映射	8a
matching	['mætʃɪŋ]	adj.相配的；一致的；相称的 n.匹配	9a
mathematical	[‚mæθə'mætɪkl]	adj.数学的；精确的	6a
measure	['meʒə]	n.测量；措施；程度 vt.测量，估量	10b
mechanically	[mə'kænɪkəlɪ]	adv.机械地	4a
mental	['mentl]	adj.精神的，思想的，心理的；智慧的	2b
merge	[mɜːdʒ]	v.合并，并入，融合	4a
microphone	['maɪkrəfəʊn]	n.麦克风，话筒	9a
microscope	['maɪkrəskəʊp]	n.显微镜	7b
middleware	['mɪdlweə]	n.中间设备，中间件	4a

续表

单 词	音 标	意 义	课次
millisecond	['mɪlɪsekənd]	n.毫秒	10b
mimic	['mɪmɪk]	vt.模仿,模拟 adj.模仿的	6a
minimize	['mɪnɪmaɪz]	vt.把……减至最低数量[程度],最小化	1a
minimum	['mɪnɪməm]	n.最小量;极小值	3a
miracle	['mɪrəkl]	n.奇迹,令人惊奇的人(或事)	9b
mirror	['mɪrə]	vt.反映,反射	6a
misstep	[ˌmɪs'step]	n.错误;失策;失足	5b
misuse	[ˌmɪs'juːz]	vt.使用……不当;滥用	3b
mobile	['məʊbaɪl]	adj.可移动的 n.手机	1b
mobility	[məʊ'bɪlɪtɪ]	n.流动性;移动性;机动性	10b
model	['mɒdl]	n.模型;模式;典型	1b
modernize	['mɒdənaɪz]	v.使现代化	1b
module	['mɒdjuːl]	n.模块	4b
monitor	['mɒnɪtə]	vt.监督,监控;测定 n.显示屏,显示器	5a
monumental	[ˌmɒnjʊ'mentl]	adj.重要的;非常大的	10b
moral	['mɒrəl]	adj.道德的	2b
multimedia	[ˌmʌltɪ'miːdɪə]	n.多媒体	4b
multitask	[ˌmʌltɪ'tɑːsk]	n.多任务	9a
multitude	['mʌltɪtjuːd]	n.大量,许多	6a
mutable	['mjuːtəbl]	adj.可变的,易变的	4b
nadir	['neɪdɪə]	n.最低点	2b
namespace	['neɪmspeɪs]	n.名空间	4b
narrow	['nærəʊ]	adj.狭隘的,狭窄的	1a
naturally	['nætʃərəlɪ]	adv.自然地,顺理成章地;合理地	7b
navigate	['nævɪgeɪt]	v.导航	5b
neighboring	['neɪbərɪŋ]	adj.邻近的	2a
nervous	['nɜːvəs]	adj.神经系统的	6b
neuron	['njʊərɒn]	n.神经元,神经细胞	8a
neutral	['njuːtrəl]	adj.中立的	1a
niche	[nɪtʃ]	n.商机	10a
nimble	['nɪmbl]	adj.灵活的;敏捷的	5b
node	[nəʊd]	n.节点	2a
noise	[nɔɪz]	n.噪声,杂音	8b
nonlinear	['nɒn'lɪnɪə]	adj.非线性的	8a
normalization	[ˌnɔːməlaɪ'zeɪʃn]	n.规范化,正常化,标准化	9a
novelty	['nɒvltɪ]	n.新奇,新奇的事物	7a
nuance	['njuːɑːns]	n.细微差别	10b
numerically	[njuː'merɪklɪ]	adv.用数表示地,数字化地	2a
object	['ɒbdʒɪkt]	n.对象;物体;目标	6a
obscure	[əb'skjʊə]	adj.不清楚的;隐蔽的 vt.使……模糊不清;掩盖,隐藏	6b

续表

单　　词	音　　标	意　　义	课次
observation	[ˌɒbzə'veɪʃn]	n.观察,观察力	1a
obstacle	['ɒbstəkl]	n.障碍,障碍物	1b
obstruction	[əb'strʌkʃn]	n.阻塞,阻碍,受阻	1a
offline	[ˌɒf'laɪn]	adj.未连线的;未联机的;脱机的;离线的 adv.未连线地;未联机地;脱机地;离线地	3b
off-the-shelf	[ɒf-ðə-ʃelf]	adj.现成的,买来不用改就用的	6a
omit	[əu'mɪt]	vt.省略;删掉	3a
on-demand	[ˌɒndɪ'mɑːnd]	adj.按需的	10b
online	['ɒn'laɪn]	adj.在线的;联网的;联机的	5b
ontological	[ˌɒntə'lɒdʒɪkl]	adj.存在论的,本体论的,实体论的	5a
ontology	[ɒn'tɒlədʒɪ]	n.本体,存在;实体论	1a
opaque	[əu'peɪk]	adj.不透明的;含糊的 n.不透明	1a
openness	['əupənnɪs]	n.开放,公开	6b
operate	['ɒpəreɪt]	v.运转;操作;经营;管理	3b
operator	['ɒpəreɪtə]	n.运算符	2a
opponent	[ə'pəunənt]	n.对手	1a
opportunity	[ˌɒpə'tjuːnɪtɪ]	n.机会	5b
optimal	['ɒptɪməl]	adj.最佳的,最优的;最理想的	7a
optimality	[ɒptɪ'mælɪtɪ]	n.最优性;最佳性	2a
optimise	['ɒptɪmaɪz]	vt.使最优化	10a
optional	['ɒpʃənl]	adj.可选择的	4b
organ	['ɔːgən]	n.器官;元件	7a
originality	[ə,rɪdʒə'nælɪtɪ]	n.独创性,创造性	2b
outperform	[ˌautpə'fɔːm]	vt.做得比……更好,胜过	7b
outreach	['autriːtʃ]	n.扩大服务范围 adj.扩大服务的	9a
overcome	[ˌəuvə'kʌm]	v.战胜,克服;压倒	2b
overwhelming	[ˌəuvə'welmɪŋ]	adj.势不可挡的,压倒一切的	1a
painstaking	['peɪnzteɪkɪŋ]	adj.辛苦的;苦干的 n.苦干,刻苦;勤勉;辛苦	2b
pair	[peə]	n.一对,一副 v.(使……)成对,(使……)成双	8b
parabolic	[ˌpærə'bɒlɪk]	adj.抛物线的	8b
paradigm	['pærədaɪm]	n.范例	4a
parallelism	['pærəlelɪzəm]	n.平行;对应,类似	8a
parameter	[pə'ræmɪtə]	n.参数	5a
parentheses	[pə'renθəsiːz]	n.圆括号	4b
parse	[pɑːz]	vt.从语法上描述或分析(词句等)	9a
particular	[pə'tɪkjulə]	adj.特别的;详细的;独有的 n.特色,特点	1a
particularity	[pə,tɪkju'lærɪtɪ]	n.特性	3a
path	[pɑːθ]	n.路径	7a
pattern	['pætn]	n.模式 vt.模仿	8a

单 词	音 标	意 义	课次
pave	[peɪv]	vt.铺设；为……铺平道路；安排	9b
pedestrian	[pə'destrɪən]	n.步行者,行人 adj.徒步的	1a
penalty	['penəltɪ]	n.惩罚,刑罚	10b
permission	[pə'mɪʃn]	n.允许,批准,认可	10b
pertinent	['pɜːtɪnənt]	adj.有关的,相干的；恰当的	6a
pervasive	[pə'veɪsɪv]	adj.普遍的；扩大的 adv.无处不在地；遍布地 n.无处不在；遍布	3b
phenomenal	[fə'nɒmɪnl]	adj.显著的	9a
philosophy	[fɪ'lɒsəfɪ]	n.哲学	10a
pick-up	['pɪkʌp]	n.提取	10b
pillar	['pɪlə]	n.台柱,顶梁柱	10b
pilot	['paɪlət]	n.引航员；向导 vt.驾驶	1a
planetary	['plænətrɪ]	adj.行星的	2b
plateau	['plætəʊ]	n.平稳时期,稳定水平；停滞期 v.达到平稳状态；进入停滞期	7b
platform	['plætfɔːm]	n.平台	9a
playbook	['pleɪbʊk]	n.剧本	5b
plethora	['pleθərə]	n.过多,过剩	10b
pointer	['pɔɪntə]	n.指针	4a
polish	['pɒlɪʃ]	n.优美,优雅,精良	5b
pool	[puːl]	n.水池	2a
position	[pə'zɪʃn]	n.位置,方位；态度；状态 vt.安置；把……放在适当位置；给……定位	2a
possess	[pə'zes]	vt.拥有；掌握,懂得	8a
possibility	[ˌpɒsə'bɪlɪtɪ]	n.可能,可能性	10b
potential	[pə'tenʃl]	adj.潜在的,有可能的	6a
pothole	['pɒthəʊl]	n.坑	6b
preamble	[prɪ'æmbl]	n.序；绪言	3a
precious	['preʃəs]	adj.珍贵的,贵重的	10a
precisely	[prɪ'saɪslɪ]	adv.精确地；恰好地	3a
precision	[prɪ'sɪʒn]	n.精确度,准确(性) adj.精确的,准确的	2b
predefine	['priːdɪ'faɪn]	vt.预先确定；预定义	8a
predict	[prɪ'dɪkt]	vt.预言,预测,预示,预告	1b
prediction	[prɪ'dɪkʃn]	n.预测,预报；预言	1a
predisposition	[ˌpriːdɪspə'zɪʃn]	n.倾向,素质	5a
preexist	['priːɪg'zɪst]	v.先前存在,预先存在的	6a
preference	['prefrəns]	n.优先权；偏爱	10a
prefix	['priːfɪks]	n.前缀	4b
preprogrammed	[pre'prəʊgræmd]	adj.预编程序的	6b
preset	[ˌpriː'set]	vt.预设,预先布置；事先安排	6b
pretrained	[priː'treɪnd]	adj.预训练的	7b

续表

单 词	音 标	意 义	课次
prevent	[prɪˈvent]	vt.防止,预防;阻碍;阻止 vi.阻止	3a
primitive	[ˈprɪmətɪv]	adj.原始的	8a
principle	[ˈprɪnsɪpl]	n.原则,原理	10a
privacy	[ˈprɪvəsɪ]	n.隐私,秘密	3b
probability	[ˌprɒbəˈbɪlɪtɪ]	n.概率;可能性,或然性	6a
procedural	[prəˈsiːdʒərəl]	adj.程序上的	4a
procedure	[prəˈsiːdʒə]	n.程序;过程,步骤	9a
process	[ˈprəʊses]	n.过程	1b
process	[ˈprəʊses]	n.过程 vt.加工;处理	9a
productivity	[ˌprɒdʌkˈtɪvɪtɪ]	n.生产率,生产力	5a
programmatic	[ˌprəʊɡrəˈmætɪk]	adj.按计划的,程序的	6a
programmer	[ˈprəʊɡræmə]	n.程序员	4b
progressively	[prəˈɡresɪvlɪ]	adv.日益增加地;逐步	6b
promote	[prəˈməʊt]	vt.促进,推进	5a
propose	[prəˈpəʊz]	vt.提议,建议;打算,计划 vi.做出计划,打算	10b
prospective	[prəˈspektɪv]	adj.预期的;可能的;有希望的	2a
prove	[pruːv]	vt.证明,证实;显示 vi.显示出,证明是	1b
provision	[prəˈvɪʒn]	n.设备;供应	10b
pseudocode	[ˈsjuːdəʊkəʊd]	n.伪代码	3a
pseudorandom	[psjuːdəʊˈrændəm]	adj.伪随机的	4b
psychology	[saɪˈkɒlədʒɪ]	n.心理学;心理特点;心理状态	1a
psychosocial	[saɪkəʊˈsəʊʃəl]	adj.社会心理的	6a
query	[ˈkwɪərɪ]	v.查询	3a
queue	[kjuː]	n.队列	2a
random	[ˈrændəm]	adj.任意的;随机的 n.随意;偶然的行动	7a
randomly	[ˈrændəmlɪ]	adv.随机地,随便地	8b
rating	[ˈreɪtɪŋ]	n.等级;评估,评价	5b
reactive	[rɪˈæktɪv]	adj.反应的	1a
readability	[ˌriːdəˈbɪlətɪ]	n.易读,可读性	4b
reader	[ˈriːdə]	n.读卡机	4a
ready-to-use	[ˈredɪ-tʊ-juːz]	adj.即用的,随时可用的	6a
realistic	[ˌriːəˈlɪstɪk]	adj.逼真的;栩栩如生的	8a
realm	[relm]	n.领域,范围	1a
reasoning	[ˈriːzənɪŋ]	n.推理,论证 v.推理,思考;争辩;说服 adj.推理的	1a
reassess	[ˌriːəˈses]	v.再估价,再评价	10a
recall	[rɪˈkɔːl]	vt.回调,再次调用	5a
receive	[rɪˈsiːv]	v.收到;接到	5a
recognise	[ˈrekəɡnaɪz]	vt.认出,识别出某人[某事物];认可,承认	10b
recognition	[ˌrekəɡˈnɪʃn]	n.识别,认识	1b

续表

单　　词	音　　标	意　　义	课次
recommend	[ˌrekəˈmend]	v.推荐；建议	6a
recommendation	[ˌrekəmenˈdeɪʃn]	n.推荐；建议	2b
record	[ˈrekɔːd]	n.记录	6a
recreate	[ˌriːkriˈeɪt]	v.重现；重建；再创造	2b
recurrent	[rɪˈkʌrənt]	adj.递归的；周期性,经常发生的；循环的	8a
recursion	[rɪˈkəːʃn]	n.递归,递推	2a
reexpansion	[riːɪkˈspænʃn]	n.再扩展	3a
reflect	[rɪˈflekt]	v.反映,反射；考虑	3b
regard	[rɪˈɡɑːd]	vt.认为；注视；涉及	5a
register	[ˈredʒɪstə]	n.&vt.登记,注册	10b
registration	[ˌredʒɪˈstreɪʃn]	n.登记,注册	5b
regression	[rɪˈɡreʃn]	n.回归	8b
regular	[ˈreɡjʊlə]	adv.定期地；经常地	5b
regulation	[ˌreɡjʊˈleɪʃn]	n.规章,规则 adj.规定的	1a
reliability	[rɪˌlaɪəˈbɪlɪti]	n.可靠性	4a
reliably	[rɪˈlaɪəbli]	adv.可靠地,确实地	6a
remove	[rɪˈmuːv]	vt.删除,去除	3a
repair	[rɪˈpeə]	vt.修理；恢复 n.修理；修理工作；维修状态	2b
repeat	[rɪˈpiːt]	v.重复	2a
repetitive	[rɪˈpetɪtɪv]	adj.重复的,啰唆的	1a
replicate	[ˈreplɪkeɪt]	vt.复制 adj.复制的	6a
repository	[rɪˈpɒzətri]	n.仓库；贮藏室	5a
represent	[ˌreprɪˈzent]	vt.代表,表现	2a
representation	[ˌreprɪzenˈteɪʃn]	n.表现；表现……的事物	5a
representative	[ˌreprɪˈzentətɪv]	n.代表 adj.典型的；有代表性的	9a
reputational	[ˌrepjuːˈteɪʃənl]	n.声誉	3b
researcher	[rɪˈsəːtʃə]	n.研究员,调查者	
resemble	[rɪˈzembl]	vt.与……相像,类似于	3a
reset	[ˌriːˈset]	vt.重置；重排；重新安装	3b
resist	[rɪˈzɪst]	v.抵抗,抗拒,抵制	10a
resolution	[ˌrezəˈluːʃn]	n.解决；坚决；分辨率	5b
response	[rɪˈspɒns]	n.响应,反应；回答,答复	6a
responsible	[rɪˈspɒnsəbl]	adj.尽责的；承担责任；负有责任的；	6b
responsibly	[rɪˈspɒnsəbli]	adv.负责地,有责任感地	10a
restrict	[rɪˈstrɪkt]	vt.限制,限定；约束	3a
restriction	[rɪˈstrɪkʃn]	n.限制,限定	3a
return	[rɪˈtəːn]	v.返回,回来；退还；重现	2a
reversed	[rɪˈvəːst]	v.翻转,颠倒	4b
reward	[rɪˈwɔːd]	n.奖赏；报酬；赏金；酬金 vt.奖赏；酬谢	7a
rigorous	[ˈrɪɡərəs]	adj.严密的；缜密的；严格的	9b
risk	[rɪsk]	n.危险,冒险 vt.冒……的危险	3b

续表

单 词	音 标	意 义	课次
roadway	['rəʊdweɪ]	n.路面,道路;车道	10b
robot	['rəʊbɒt]	n.机器人;遥控装置;自动机	2b
robotics	[rəʊ'bɒtɪks]	n.机器人技术	9b
role	[rəʊl]	n.角色;作用;地位	6a
rounding	['raʊndɪŋ]	n.舍入,取整	4b
route	[ru:t]	n.路径,途径	2b
routine	[ru:'ti:n]	n.常规,例行程序 adj.例行的,常规的,日常的	10a
rule	[ru:l]	n.规则,规定;统治,支配 v.控制,支配	1a
rule	[ru:l]	n.规则,规定;统治,支配 vi.控制,支配	3a
ruling	['ru:lɪŋ]	adj.统治的;支配的;管辖的 n.统治;支配	2b
runtime	['rʌntaɪm]	n.运行时间,运行期	8a
sample	['sɑ:mpl]	n.样本,样品 vt.取……的样品;抽样调查	8b
satellite	['sætəlaɪt]	n.卫星,人造卫星	7b
satisfaction	[,sætɪs'fækʃn]	n.满足,满意	5b
satisfy	['sætɪsfaɪ]	v.使满意,满足	9a
savvy	['sævɪ]	n.机智;头脑 adj.有见识的	6b
scarce	[skeəs]	adj.罕见的	6a
scenario	[sə'nɑ:rɪəʊ]	n.设想;可能发生的情况;剧情梗概	7a
schedule	['ʃedju:l]	n.进度表,明细表;预定计划 vt.排定,安排	10a
schema	['ski:mə]	n.模式;概要,计划	5a
score	[skɔ:]	n.&v.得分;记分	3a
script	[skrɪpt]	n.脚本	4a
search	[sɜ:tʃ]	v.搜索,搜寻;调查 n.搜索	2a
security	[sɪ'kjʊərɪtɪ]	n.安全;保证,保护 adj.安全的,保安的,保密的	10b
self-assess	[self-ə'ses]	n.自我评估,自主评估	9b
self-awareness	[self-ə'weənɪs]	n.自我意识	1a
self-correction	[,selfkə'rekʃn]	n.自校正;自我纠错;自我改正	1a
self-decide	[self-dɪ'saɪd]	n.自主决定	9b
self-serve	['selfsɜ:v]	adj.自我服务的,自助的	5b
sense	[sens]	n.感觉 vt.感到;理解,领会	5a
sensor	['sensə]	n.传感器	9a
sentience	['senʃəns]	n.感觉性;感觉能力;知觉	5a
sentient	['sentɪənt]	adj.有感觉能力的,有知觉力的	1a
sentiment	['sentɪmənt]	n.感情,情绪;意见,观点	2b
separated	['sepəreɪtɪd]	adj.分隔的,分开的	7a
sequence	['si:kwəns]	n.序列;顺序;连续 vt.使按顺序排列	2a
shallow	['ʃæləʊ]	adj.浅的,肤浅的	7b
shape	[ʃeɪp]	n.形状;模型 vi.使成形;形成	7b
shockingly	['ʃɒkɪŋlɪ]	adj.令人震惊地,极度地	8a

单 词	音 标	意 义	课次
shortcoming	[ˈʃɔːtkʌmɪŋ]	n.短处,缺点	9b
signal	[ˈsɪgnəl]	n.信号 vt.向……发信号 vi.发信号	5a
signature	[ˈsɪgnɪtʃə]	n.签名;署名;识别标志	1a
significant	[sɪgˈfɪkənt]	adj.重要的;显著的;有重大意义的	6a
similarity	[ˌsɪmɪˈlærɪti]	n.相像性,相仿性,类似性	1a
simplify	[ˈsɪmplɪfaɪ]	vt.简化,使简易	5b
simulation	[ˌsɪmjʊˈleɪʃn]	n.模仿,模拟	1a
simulation	[ˌsɪmjʊˈleɪʃn]	n.模仿,模拟	5a
simultaneously	[ˌsɪməlˈteɪnɪəsli]	adv.同时地	8a
situation	[ˌsɪtjʊˈeɪʃn]	n.(人的)情况;局面,形势,处境;位置	7a
skilful	[ˈskɪlfl]	adj.灵巧的;熟练的;技术好的	9a
skill	[skɪl]	n.技能,技巧;才能,本领	1b
smart	[smɑːt]	adj.聪明的;敏捷的	5b
smartphone	[smɑːtfəʊn]	n.智能手机	9a
software	[ˈsɒftweə]	n.软件	9a
solely	[ˈsəʊlli]	adv.唯一地;仅仅	3a
solution	[səˈluːʃn]	n.解决;答案	2a
sophisticated	[səˈfɪstɪkeɪtɪd]	adj.复杂的;精致的;富有经验的	3b
sphere	[sfɪə]	n.范围;势力范围 vt.包围,围绕	9b
splash	[splæʃ]	v.(使)溅起;引人注目 n.扑通声	9a
spot	[spɒt]	v.认出,发现	8a
sprinkler	[ˈsprɪŋklə]	n.洒水器,自动喷水灭火装置	8a
sprout	[spraʊt]	vi.发芽;抽芽 vt.使发芽;使生长	10a
stack	[ˈstæk]	n.堆栈	2a
stage	[steɪdʒ]	n.阶段	3b
stakeholder	[ˈsteɪkhəʊldə]	n.股东;利益相关者	10a
standardize	[ˈstændədaɪz]	vt.使标准化	5b
startup	[ˈstɑːtʌp]	n.新兴公司;启动	10a
state	[steɪt]	n.状态	2a
statement	[ˈsteɪtmənt]	n.语句	4b
state-of-the-art	[steɪt-əv-ðɪ-ɑːt]	adj.使用最先进技术的	5a
static	[ˈstætɪk]	adj.静止的;不变的	7a
stationary	[ˈsteɪʃənri]	adj.不动的,固定的;静止的,不变的	10b
statistical	[stəˈtɪstɪkl]	adj.统计的,统计学的	1b
steer	[stɪə]	v.驾驶;操纵,控制;引导	6b
strength	[streŋθ]	n.力量;优点,长处	6b
strict	[strɪkt]	adj.严格的;精确的;绝对的	1a
string	[strɪŋ]	n.串	4b
structural	[ˈstrʌktʃərəl]	adj.结构的,结构化	4a
stuff	[stʌf]	n.材料,原料,资料	5b
sub-discipline	[sʌb-ˈdɪsəplɪn]	n.子学科	1b

续表

单 词	音 标	意 义	课次
substance	[ˈsʌbstəns]	n.实质,内容	1b
substantial	[səbˈstænʃl]	adj.大量的;重大的;结实的,牢固的	7b
successive	[səkˈsesɪv]	adj.连续的	4b
successor	[səkˈsesə]	n.接替的人或事物;继任者	2a
suddenly	[ˈsʌdənlɪ]	adv.意外地,忽然地	10b
sufficiently	[səˈfɪʃntlɪ]	adv.足够地,充分地;十分,相当	6a
suit	[suːt]	vt.适合于	2b
suitable	[ˈsjuːtəbl]	adj.合适的,适当的	8b
supersede	[ˌsuːpəˈsiːd]	vt.取代,接替	2b
supervise	[ˈsjuːpəvaɪz]	v.监督;管理;指导	8b
supposition	[ˌsʌpəˈzɪʃn]	n.推测,猜测;假定	5a
surgery	[ˈsɜːdʒərɪ]	n.外科学,外科手术	9b
surgically	[ˈsɜːdʒɪklɪ]	adv.精确地,如外科手术般地	7b
surrogate	[ˈsʌrəgɪt]	n.代理,代表 adj.代理的 v.代理,替代	5a
surround	[səˈraund]	vt.包围,围绕	10b
surveillance	[sɜːˈveɪləns]	n.监督	3b
survey	[ˈsɜːveɪ]	vt.调查;勘测 n.调查(表)	9a
sustain	[səˈsteɪn]	vt.维持;支撑,支持	6a
symbiotic	[ˌsɪmbaɪˈɒtɪk]	adj.共生的	9b
synchronize	[ˈsɪŋkrənaɪz]	vt.使同步;使同时 vi.同时发生;共同行动	6a
synthesize	[ˈsɪnθəsaɪz]	v.综合,合成	6a
systematic	[ˌsɪstəˈmætɪk]	adj.系统的,规则的;有步骤的	6a
system-readable	[ˈsɪstəm-ˈriːdəbl]	adj.系统可读的	5b
tactile	[ˈtæktaɪl]	adj.触觉的	5a
tailored	[ˈteɪləd]	adj.定做的,特制的,专门的	4a
taxonomy	[tækˈsɒnəmɪ]	n.分类学,分类系统	1a
technology	[tekˈnɒlədʒɪ]	n.科技(总称);工业技术	1b
telecommunication	[ˌtelɪkəˌmjuːnɪˈkeɪʃn]	n.电信	8a
template	[ˈtempleɪt]	n.模板(=templet)	4a
terminate	[ˈtɜːmɪneɪt]	v.结束,使终结	2a
terrain	[təˈreɪn]	n.地面,地带	3a
text	[tekst]	n.文本	9a
textual	[ˈtekstʃuəl]	adj.文本的,正文的,原文的	6b
theory	[ˈθɪərɪ]	n.理论;原理	5a
threat	[θret]	n.威胁	3b
thrill	[θrɪl]	vt.使兴奋,使激动	9b
thrive	[θraɪv]	vi.兴盛,茁壮成长	10a
throughput	[ˈθruːpʊt]	n.吞吐量;流率	8a
tile	[taɪl]	n.片状材料,瓦片、瓷砖 vt.用瓦片、瓷砖等覆盖	3a
timeliness	[ˈtaɪmlɪnɪs]	n.及时性	6a

续表

单　　词	音　　标	意　　义	课次
time-tested	[ˈtaɪmˈtestɪd]	adj.经受时间考验的，久经试验的	1b
togetherness	[təˈgeðənəs]	n.亲密无间；和睦；团结	2b
token	[ˈtəʊkən]	n.记号 adj.作为标志的	9a
tokenization	[ˌtəʊkəaɪˈzeɪʃn]	n.词语切分	9a
toolkit	[ˈtuːlkɪt]	n.工具包，工具箱	1b
topic	[ˈtɒpɪk]	n.主题；话题，论题	5b
trace	[treɪs]	vt.跟踪，追踪；追溯	3a
trademark	[ˈtreɪdmɑːk]	n.（注册）商标	5b
traditional	[trəˈdɪʃnl]	adj.传统的；惯例的	5a
traffic	[ˈtræfɪk]	n.交通，运输量	10b
train	[treɪn]	v.训练；教育；培养	7a
trajectory	[trəˈdʒektərɪ]	n.轨道	10b
transactional	[trænˈzækʃənl]	adj.业务的，交易的	6a
transcribe	[trænˈskraɪb]	vt.转录；改编（乐曲）	1b
transformation	[ˌtrænsfəˈmeɪʃn]	n.转换；变化	8a
transformative	[ˌtrænsˈfɔːmətɪv]	adj.有改革能力的，变化的，变形的	10a
translation	[trænsˈleɪʃn]	n.翻译	7b
transparency	[trænsˈpærənsɪ]	n.透明度，透明性	10a
treatment	[ˈtriːtmənt]	n.处理；待遇，对待	6a
trivial	[ˈtrɪvɪəl]	adj.无价值的；平常的，平凡的；不重要的	6b
troubleshooting	[ˈtrʌblʃuːtɪŋ]	n.发现并修理故障	5b
trusted	[ˈtrʌstɪd]	adj.可信的，无错的	3b
tune	[tjuːn]	n.曲调；调谐 vt.调整	8a
tuple	[ˈtʌpl]	n.元组	4b
tweak	[twiːk]	vt.稍稍调整（机器、系统等）	7b
unaffected	[ˌʌnəˈfektɪd]	adj.不受影响的	2b
unavailability	[ˈʌnəˌveɪləˈbɪlɪti]	n.无效用，不适用	9a
unavoidable	[ˌʌnəˈvɔɪdəbl]	adj.不可避免的，不得已的	1a
uncertainty	[ʌnˈsɜːtntɪ]	n.无把握，不确定；不可靠	6a
unclear	[ˌʌnˈklɪə]	adj.不清楚的，不明白的，含糊不清	1a
unclutter	[ʌnˈklʌtə]	vt.使整洁，整理	4b
uncover	[ʌnˈkʌvə]	vi.发现，揭示	6b
underpin	[ˌʌndəˈpɪn]	vt.加固，支撑	1a
undoubtedly	[ʌnˈdaʊtɪdlɪ]	adv.毋庸置疑地，的确地；显然；必定	9a
unemployment	[ˌʌnɪmˈplɔɪmənt]	n.失业；失业率；失业状况	2b
unexpanded	[ˌʌnɪkˈspændɪd]	adj.未被扩大的，未展开的	2a
unexpected	[ˌʌnɪkˈspektɪd]	adj.意外的；忽然的；突然的	1a
unfamiliar	[ˌʌnfəˈmɪlɪə]	adj.不熟悉的；不常见的；陌生的；没有经验的	1a
universal	[ˌjuːnɪˈvɜːsl]	adj.普遍的，一般的，通用的	2a
universally	[ˌjuːnɪˈvɜːsəlɪ]	adv.普遍地，一般地；人人，处处	5a

续表

单 词	音 标	意 义	课次
unlabelled	[ʌn'leɪbld]	adj.未标记的	8b
unpeel	['ʌn'piːl]	v.削……的皮,剥离	10b
unrealistic	[ˌʌnrɪə'lɪstɪk]	adj.不切实际的;不现实的;空想的	1a
unsafe	[ʌn'seɪf]	adj.不安全的,危险的	7b
unstructured	[ʌn'strʌktʃəd]	adj.非结构化的,无结构的	5a
unsupervise	[ʌn'sjuːpəvaɪz]	v.无监督;无管理	8b
unvisited	[ʌn'vɪzɪtɪd]	adj.未访问的	3a
update	[ˌʌp'deɪt]	vt.更新 n.更新	7a
updation	['ʌpdeɪʃn]	n.上升	9b
upliftment	['ʌplɪftmənt]	n.提升	9b
upload	[ˌʌp'ləʊd]	vt.上传,上载	5b
usability	[ˌjuːzə'bɪlɪtɪ]	n.可用性;适用性	9a
usage	['juːsɪdʒ]	n.使用,用法	5a
utility	[juːˈtɪlətɪ]	n.功用,效用	2b
valid	['vælɪd]	adj.有效的	2a
validate	['vælɪdeɪt]	vt.确认;证实	7a
valuable	['væljʊəbl]	adj.贵重的,宝贵的;有价值的	3b
variant	['veərɪənt]	n.变体,变种,变异体;变量 adj.不同的,相异的,不一致的;多样的;变异的	3a
vehicle	['viːɪkl]	n.车辆;交通工具	1a
verification	[ˌverɪfɪ'keɪʃn]	n.核实;证实	9a
versed	[vɜːst]	adj.精通的,熟练的	5b
vertical	['vɜːtɪkl]	adj.垂直的,竖立的 n.垂直线,垂直面	7a
victorious	[vɪk'tɔːrɪəs]	adj.胜利的,得胜的	9a
virtual	['vɜːtʃʊəl]	adj.(计算机)虚拟的;实质上的,事实上的	1a
vision	['vɪʒn]	n.视觉	1a
visual	['vɪʒuəl]	adj.视觉的,看得见的	5a
voice	[vɔɪs]	n.语音	1b
vulnerability	[ˌvʌlnərə'bɪlɪtɪ]	n.弱点	3b
weigh	[weɪ]	v.权衡,考虑	2b
weight	[weɪt]	n.权重	3b
wholeheartedly	[ˌhəʊl'hɑːtɪdlɪ]	adv.全心全意地,全神贯注地;真心诚意	2b
witness	['wɪtnɪs]	vt.表示,提供……的证据	9a
workflow	['wɜːkfləʊ]	n.工作流程,工作流	7b
workforce	['wɜːkfɔːs]	n.劳动力,劳动人口	10a
worry	['wʌrɪ]	n.烦恼,忧虑;担心 v.担心,焦虑,发愁	9a

A.2 词 组 表

词 组	意 义	课次
a blend of	混合	9a
a fraction of	一小部分	9b
a kind of …	……的一种	6a
a lack of	缺乏，缺少	10b
a lot of	许多的	1b
a piece of	一块；一片；一件	5a
a range of	一系列，一些，一套	1a
a sense of …	一种……感觉	1a
a variety of	多种的	4a
a wide range of	广泛的	1b
absolute value	绝对值	6a
according to	根据，按照	2a
accustomed to	习惯于……	9a
adapt to	使适应于	3b
AI-enabled chip	人工智能芯片	9a
AI-oriented computational job	面向人工智能的计算任务	1b
all sorts of	各种各样的	4a
along with	和……一起，随着；以及；连同	9a
analog computer	模拟计算机	4a
analog-to-digital conversion	模(拟)数(字)转换	1a
anonymous function	匿名函数	4b
array index	数组下标	4b
array slicing	数组切片	4b
arrived at	到达	6b
artificial general intelligence	通用人工智能	1a
artificial neural network	人工神经网络	8a
as much as possible	尽可能	4a
as the name suggests	顾名思义	7a
as well as	也，又	1b
aspect-oriented programming	面向切面编程	4b
assembly language	汇编语言	4a
assembly line	(工厂产品的)装配线，流水线	1a
assignment statement	赋值语句	4b
at hand	在手边，在附近；即将来临	6a
autonomous driving	自主驾驶	10b
autonomous drone	自主无人机	6b
baby stroller	婴儿车	10b
backward pass	逆推法，倒推法	8a
base on	基于，建立在……上	5a

词　　组	意　　义	课次
be able to	能，会	4a
be aware of	知道	4a
be capable of	能够	9a
be categorized as…	被分类为……	1a
be compared with …	与……相比较	9a
be concatenated with …	与……连接	2a
be confused with	混淆	4a
be educated on	接受教育	10b
be conflated with …	与……混为一谈	1a
be incapable of	不能	4a
be incorporated into…	被并入……	1a
be integrated into …	统一到……中，整合到……中	4a
be known as	被认为是	4a
be opposed to	与……相对，和……相反	4a
be shaping up	正在成形	10a
be thrown at	被扔向	8a
be treated as	被当作	3a
been built upon	建立在	5a
bicycle sharing	共享自行车	10b
bidirectional search	双向搜索，双向查找	2a
black box	黑盒子，黑匣子	8a
blank slate	空白板	5a
body language	身体语言，手势语言	1b
Boltzmann machine network	玻尔兹曼机器网络	8a
boolean expression	逻辑表达式，布尔表达式	4b
boolean operator	布尔运算符，逻辑运算符	4b
branching factor	分支因子，分支系数	2a
breadth-first search	宽度优先搜索	2a
brute-force search	蛮力搜索，强力搜索	2a
business case	商业案例	1b
business forecast	业务预测	9b
business model	企业模型，商业模式	3b
business purpose	商业目的，营业目的	1a
carry out	完成，实现，执行	4a
case study	案例研究，个案研究；范例分析	6a
case-based reasoning	实例推理(法)，基于案例推理	6a
caught up with	追上，赶上	9a
child node	子节点	2a
city planner	城市规划师	10b
cloud computing	云计算	7b
cognitive model	认知模型	5a

续表

词　　组	意　　义	课次
coined in	发明于	1b
come up with	追赶上；提出；想出；设法拿出	9b
come with	伴随……发生，与……一起供给	4a
command line interpreter	命令行解释程序	4b
common phrase	常见短语，常用短语	9a
computational model	计算模型	8a
computer complex	计算装置	4a
computer game	计算机游戏程序	4a
computer science	计算机科学	1a
computer-based information system	基于计算机的信息系统	5a
computer-controlled bot	计算机控制的机器人	2b
computing power	计算能力	1b
concerned with	涉及；与……有关	7a
conditional expression	条件表达式	4b
conditional instruction	条件指令	4a
confidence scoring schema	置信评分模式	1a
consist of	由……组成；包括	3a
consisting of …	由……组成	1b
consumption pattern	消费模式	10a
control flow	控制流	4b
convolutional layer	卷积层	7b
convolutional neural network	卷积神经网络	8a
crime group	犯罪集团	3b
critical period	关键时期	9b
crystal ball	（占卜用的）水晶球，预言未来的方法	10a
curly braces	大括号，花括号，大括弧	4b
customer service	客户服务	1b
customer-facing application	面向客户的应用	1b
data analytic	数据分析	9b
data mining	数据挖掘	8a
data oriented	数据导向	7a
data privacy	数据保密	9a
data set	数据集	1b
data storehouse	数据仓库	9b
data structure	数据结构	2a
dead-end node	死角节点	3a
decision making process	决策过程	8a
decision-making process	决策程序，决策过程	6b
deep learning	深度学习	7b
deep learning algorithm	深度学习算法	1a
deep neural network	深度神经网络，深层神经网络	8a

词　　组	意　　义	课次
Density Based Spatial Clustering Application	基于密度的空间聚类应用	8b
density-based clustering	基于密度的聚类方法	8b
depth-first search	深度优先搜索	2a
derived from …	来源于……	3b
design by contract	契约设计	4b
design philosophy	设计原理	4b
desired value	期望值	8a
development environment	开发环境	4b
device driver	设备驱动程序	4a
diagnostic system	诊断系统	6a
diagnostic tool	诊断工具	4a
difference engine	差分机	4a
digital assistant	数字助理	2b
digital communication	数字通信系统	6a
digital format	数字格式	6a
digital revolution	数字革命	9a
digital signal	数字信号	1a
distributed system	分布式的计算机系统	4a
divide … into …	把……分成……	4a
divide up	分割	3a
double quote mark	双引号	4b
driverless car	无人驾驶汽车	7b
dynamic programming language	动态编程语言	4b
economic growth	经济增长	10a
electric scooter	电动车,电动踏板车,电动代步车	10b
embedded system	嵌入式系统	4a
end-to-end learning	端到端学习	7b
enterprise intelligence	企业智能	1b
escape character	转义字符	4b
ethical principle	道德原则	10a
evaluation function	评价函数	3a
expert system	专家系统	1a
face recognition	面貌识别	3b
facial recognition	人脸识别,面孔识别,面部识别	6b
fast clustering algorithm	快速聚类算法	8b
feature extraction	特征提取	7b
feature extractor	特征提取器	7b
feedforward neural network	前馈神经网络	8a
figure out	弄明白;解决;想出	5b
financial institution	金融机构	1a
financial risk	财务风险	3b

续表

词　　组	意　　义	课次
fix bug	修复错误	3b
flash memory	闪存	3b
formulate business strategies	制定商业策略,制定业务策略	9b
forward pass	正推法	8a
fraud detection	欺诈检测	8a
freed up	被释放	10b
fuzzy logic	模糊逻辑	6a
gauge emotion	判定情绪	9a
general problem-solving rule	普通问题解决规则	3a
generator expression	生成器表达式	4b
gesture recognition	手势识别	8b
goal state	目标状态	2a
graph-based search algorithm	基于图的搜索算法	3a
greedy best first search	贪婪最佳优先搜索	2a
hands-free speaker	免提扬声器	7b
have the advantage of	胜过	4b
heavy machinery	重型机械	7b
hidden layer	隐藏层	7b
hidden pattern	隐藏模式,隐含模式	8b
high-dimensional data set	高维数据集	7b
high-level programming languages	高级编程语言	4a
hill-climbing search	爬山算法	2a
Hopfield network	霍普菲尔德网络	8a
hostile environment	有害环境	2b
hosting environment	托管环境	4b
human behavior	人类行为	9b
human intelligence	人类智能	1a
human resource	人力资源	2b
human sentience	人类感觉	5a
hypertext manipulation system	超文本操作系统	5a
image recognition	图像识别	1a
in a flash	立刻,一瞬间	10b
in a manner	在某种意义上	4b
in contrast to	和……形成对比,和……形成对照	4a
in tandem with	同……串联,同……合作	5a
in total	整个地(= as a whole)	4a
inference engine	推理机,推理引擎	6a
information travel	信息传播,信息传输	8a
informational infrastructure	信息基础设施	9a
informed (heuristic) search	启发式搜索	2a
initial state	起始状态,初态	2a

续表

词　　组	意　　义	课次
instant decision	即时决策	9b
integrated development environment（IDE）	集成开发环境	4a
intelligent cloud	智能云	9b
intelligent cloud computing	智能云计算	9b
intelligent reasoning	智能推理	5a
intelligent tutoring system	智能辅导系统	5a
interact with…	与…相互作用,与……相互影响,与……相互配合	2b
interface engine	接口引擎	5a
internet industry	互联网产业	9a
interplanetary space	太空	2b
iterative deepening depth-first search	迭代深化深度优先搜索	2a
knowledge acquisition	知识收集,知识获取	5a
knowledge database	知识数据库	6a
knowledge engineer	知识工程师	6a
knowledge management	知识管理	5b
knowledge modeling	知识模型化,知识建模	6a
knowledge repository	知识仓库	5a
knowledge representation	知识表达,知识表现	5a
knowledge set	知识集	7a
knowledge-based information system	知识库信息系统	6a
knowledge-based system	基于知识的系统	5a
labeled data	标记数据,标签化数据	7b
labour market	劳动力市场	10a
lambda expression	λ表达式	4b
lane detection	车道检测	8a
language recognition	语言识别	8a
leaf node	叶节点	2a
learning algorithm	学习算法	8a
learning phase	学习阶段	9b
leave comment	留下评论	5b
liberal art	文科	10a
library function	库函数	4b
linear regression	线性回归	8b
lines of code	代码行	4b
list comprehension	列表解析,列表推导	4b
local beam search	局部集束搜索	2a
local search algorithm	局部搜索算法	2a
logic programming	逻辑编程	4b
machine language	机器语言	4a

续表

词　　组	意　　义	课次
machine vision	机器视觉	1a
machine-based image processing	基于机器的图像处理	1a
make a distinction between…	对……加以区别	4b
make sure	确保	5b
markup language	标识语言	9a
mathematical argument	数学论证	6a
matrix multiplication	矩阵乘法	4b
media attention	媒体关注度	1b
medical device	医疗设备	7b
medical image analysis	医学图像分析	1a
memory management	内存管理	4b
mimic human reasoning	模仿人类推理	6a
mobile application	移动应用	1b
movie box office	电影票房	8a
narrow down	(使)变窄,(使)减少,(使)缩小	5b
nascent stage	初级阶段	9b
National Science and Technology Council	国家科学技术委员会	1a
natural language	自然语言	4a
natural language generation	自然语言生成	1b
nervous system	神经系统	6b
neural layer	神经层	8a
neural net	神经网络	8a
neural unit	神经单元	8a
nitty gritty	本质;实质;基本事实	5b
nonlinear feature	非线性特征	8a
nonlinear function	非线性函数	8a
not-so-distant future	不远的将来	1a
object code	结果代码	4a
object identity	对象标识	4b
off-side rule	越位规则	4b
one after another	接连地	4a
optimum performance	最佳性能	6a
ordered sequence	有序序列	4a
over and over again	一再地;来回来去;再三再四	5b
parallel architecture	并行体系结构	7b
parent page	父页面	3a
parking management	存车管理,停车管理	10b
particular case	特别情况,特例	4b
pattern recognition	模式识别	8a
personal assistant	个人助手,个人助理	9b
personal data privacy	个人数据隐私	3b

续表

词　　组	意　　义	课次
personalized experience	个性化体验	9a
physical object	物理物体	6b
physical property	物理属性	2a
platform software	平台软件	4a
post-development cost	开发后成本	6a
pour into	不断地涌进；注入；倾注	5b
power up	加电，使启动	3b
predicate logic	谓词逻辑	6a
predictive analytic	预测分析	1a
pre-existing network	现存网络	7b
preprogrammed rule	预编程序规则	6b
priority queue	优先队列	2a
problem solving	解决问题	2b
problem space	问题空间	2a
problem-solving procedure	问题解决过程	5a
pseudorandom number generator	伪随机数产生器	4b
public surveillance	群众监督	3b
public transport	公共交通，公共交通工具	10b
pull into	开进，开向路边	10b
pure heuristic search	纯启发式搜索	2a
pure science	（区别于应用科学的）纯科学	8a
rational thinking	理性思维	2b
real-time data	实时数据	3b
recognizing face	识别人脸	3b
recurrent neural network	递归神经网络	8a
refer to	涉及；指的是；适用于；参考	3a
regression model	回归模型	8b
regular expression	正则表达式	4b
reinforcement learning	强化学习	1a
resource-constrained edge device	资源受限的边缘设备	8a
restricted to	仅限于	9a
retail sector	零售部门	9a
ride-share vehicle	共享汽车，拼车	10b
road sign	交通标志，路标	7a
root node	根节点	2a
rule set	规则集	6a
safety-critical application	安全关键应用	7b
sample data	样本数据	8b
science fiction movies	科幻电影	9a
search algorithm	搜索算法	2a
search engine	搜索引擎	5a

续表

词　　组	意　　义	课次
search term	搜索词	5b
self-contained variable	自包含变量	6a
self-driving car	自动驾驶汽车	1a
self-learning mechanism	自学习机制	9b
semantic network	语义网络	6a
sentient system	感觉系统	1a
sentiment analysis	情感分析	9a
sentiment analysis	情感分析,倾向性分析	1a
separate from	分离,分开	4a
serve as	充当,担任	6a
session state retention	会话状态保留	4b
set out to	打算,着手	8a
shallow copy	浅拷贝	4b
shallow learning	浅层学习	7b
shortest path	最短路径	2a
simulated annealing	模拟退火	2a
single quote mark	单引号	4b
situation oriented	情境导向	9b
sliding-tiles game	滑动拼图游戏	2a
smart city	智慧城市	10b
smart data	智能数据	9b
smart home manager	智能管家	1b
smart office	智能办公室	9a
soak up	吸收,(使)充满	6b
social setting	社会环境,社会场景,社会情境	1a
societal benefit	社会效益	10a
societal concern	社会关注	10a
societal rule	社会规则	10a
software engineering	软件工程	4a
software reliability	软件可靠性	4a
soil condition	土壤情况	10a
source code	源编码,原代码,源程序	4a
space complexity	空间复杂度	2a
space exploration	空间探索	2b
spam detection	垃圾邮件检测	1a
special case	特殊情况	4b
speech recognition	语音识别	1a
speed up	加速;增速	9a
spelling mistake	拼写错误	9a
standard library	标准库	4b
stock market	股票市场;股票买卖;股票行情	8a

续表

词 组	意 义	课次
stop sign	停车标志	7b
storage location	存储位置,存储单元	4b
straight line	直线	7a
street sign	街道标志	3b
string interpolation	字符串插值	4b
string literal	字符串字面量	4b
strong AI	强人工智能	1a
structured data	结构化数据	9a
structured programming	结构化编程	4b
super human	超人	6b
supervised learning	有监督学习	1a
supplemental material	辅助材料	6a
surrounding tissue	外围组织,周围组织	2b
symbiotic relation	共生关系	9b
symbolic address	符号地址	4b
system software	系统软件	4a
systematic testing	系统测试	6a
tabula rasa	白板(拉丁语)	5a
take a look into	看一看	7a
talking point	话题;论题;论据	1a
tech sector	技术部门	10a
text analytic	文本分析	1b
text editor	文本编辑器	4a
text translation	文本翻译	1a
third-party tool	第三方工具	4b
tile game	智力拼图	2a
time complexity	时间复杂度	2a
time-consuming task	耗时任务	2b
touch recognition	触摸识别	1b
traced backward	追溯	3a
traffic congestion	交通拥堵	10b
traffic incident	交通事故,交通事件	10b
traffic signal	交通信号;红绿灯	7a
training cost	培训费用	5b
transfer learning	迁移学习	7b
translation tool	翻译工具	6b
travelling salesman problem	旅行商问题	2a
trend to	趋向,趋于	9a
triple-quoted string	三重引号字符串	4b
troubleshooting guide	故障排除指南	5b
type size	字号	5b

续表

词 组	意 义	课次
typographical error	印刷错误,排字错误	9a
uniform cost search	等代价搜索,一致代价搜索	2a
unlabelled data	未标记的数据	8b
unstructured data	非结构化数据	5a
unsupervised learning	无监督学习	1a
use case	用例	3b
video game	计算机视频游戏,电视游戏	4a
video stream	视频流	3b
video surveillance	视频监控	10b
virtual health assistant	虚拟健康助理	1a
virtual personal assistant	虚拟个人助理	1a
voice assistant	语音助理	9a
voice command	语音命令	9a
voice recognition	声音识别,语音识别	8a
weak AI	弱人工智能	1a
weather forecast	天气预报	9a
web crawler	网络爬虫	3a
web page	网页	3a
web scraping	网络爬虫,网络数据抓取	4b
wireless communication	无线通信	7a
worst case	最坏情况,最坏条件	2a
weave together	编织在一起	6a

A.3 缩 写 表

缩 写	意 义	课次
AI (Artificial Intelligence)	人工智能	1a
AIaaS (Artificial Intelligence as a Service)	人工智能即服务	1a
API (Application Programming Interface)	应用程序编程接口	1b
CASE (Computer Aided Software Engineering)	计算机辅助软件工程	5a
CCTV (Closed-Circuit TeleVision)	闭路电视	10b
CGI (Computer-Generated Imagery)	计算机产生的图像	8a
CNN (Convolutional Neural Network)	卷积神经网络	7b
CRM (Customer Relationship Management)	客户关系管理	1a
FAQ (Frequently Asked Questions)	常见问题	5b
FIFO (First Input First Output)	先入先出	2a
FR (Facial Recognition)	人脸识别,面孔识别	9a
GDP(Gross Domestic Product)	国内生产总值	10a
GDPR (General Data Protection Regulation)	普通数据保护条例	1a
GPS (Global Positioning System)	全球定位系统	2b

续表

缩　写	意　义	课次
GPU（Graphics Processing Unit）	图形处理单元,图形处理器	1b
HR（Human Resource）	人力资源	5b
HTTP（HyperText Transfer Protocol）	超文本传输协议	4b
IDC（International Data Corporation）	国际数据公司	1b
IoT（Internet of Things）	物联网	9b
IT（Information Technology）	信息技术	5b
KBS（Knowledge-Based System）	知识库系统	5a
KNN（K-Nearest Neighbors）	K-最近邻算法	8b
LBPH（Local Binary Pattern Histogram）	局部二值模式直方图	9a
LIFO（Last In First Out）	后进先出法	2a
MIME（Multipurpose Internet Mail Extensions）	多用途互联网邮件扩展	4b
NASA（National Aeronautics and Space Administration）	美国航空航天局	1a
NLP（Natural Language Processing）	自然语言处理	1b
NLP（Natural Language Processing）	自然语言处理	1a
NLU（Natural Language Understanding）	自然语言理解	9a
NOP（No Operation）	无操作	4b
R&D（research and development）	科学研究与开发	3b
RAII（Resource Acquisition Is Initialization）	资源获得即初始化	4b
REPL（read-eval-print loop）	读取-求值-打印循环	4b
RPA（Robotic Process Automation）	机器人流程自动化	1a
SoCs（System-on-Chips）	片载系统	3b
SVM（Support Vector Machines）	支持向量机	7b
TB（TeraByte）	兆字节	7a
VA（Virtual Assistant）	虚拟助手	9a

图书资源支持

感谢您一直以来对清华版图书的支持和爱护。为了配合本书的使用，本书提供配套的资源，有需求的读者请扫描下方的"书圈"微信公众号二维码，在图书专区下载，也可以拨打电话或发送电子邮件咨询。

如果您在使用本书的过程中遇到了什么问题，或者有相关图书出版计划，也请您发邮件告诉我们，以便我们更好地为您服务。

我们的联系方式：

地　　址：北京市海淀区双清路学研大厦 A 座 714

邮　　编：100084

电　　话：010-83470236　　010-83470237

客服邮箱：2301891038@qq.com

QQ：2301891038（请写明您的单位和姓名）

资源下载：关注公众号"书圈"下载配套资源。

资源下载、样书申请

书圈

获取最新书目

观看课程直播